Quantum Computing for Everyone

Quantum Computing for Everyone

Chris Bernhardt

The MIT Press
Cambridge, Massachusetts
London, England

First MIT Press paperback edition, 2020
© 2019 Massachusetts Institute of Technology

This book was set in ITC Stone Sans Std and ITC Stone Serif Std by Toppan Best-set Premedia Limited. Printed and bound in the United States of America.

Library of Congress Cataloging-in-Publication Data

Names: Bernhardt, Chris, author.
Title: Quantum computing for everyone / Chris Bernhardt.
Description: Cambridge, MA : The MIT Press, [2019] | Includes bibliographical
   references and index.
Identifiers: LCCN 2018018398 | ISBN 9780262039253 (hardcover : alk. paper)—
   9780262539531 (paperback)
Subjects: LCSH: Quantum computing—Popular works.
Classification: LCC QA76.889 .B47 2019 | DDC 006.3/843—dc23 LC record
   available at https://lccn.loc.gov/2018018398

10   9   8   7

To Henryka

Contents

Contents

Acknowledgments

I am very grateful to a number of people for their help with this book. Matt Coleman, Steve LeMay, Dan Ryan, Chris Staecker, and three anonymous reviewers read through various drafts with great care. Their suggestions and corrections have improved this book beyond measure. I also thank Marie Lee and her team at MIT Press for all of their support and work in turning a rough proposal into this book.

Introduction

The aim of this book is to give an introduction to quantum computing that anyone who is comfortable with high school mathematics and is willing to put in a little work can understand. We will study qubits, entanglement, quantum teleportation, and quantum algorithms, among other quantum-related topics. The goal is not to give some vague idea of these concepts but to make them crystal clear.

Quantum computing is often in the news: China teleported a qubit from earth to a satellite; Shor's algorithm has put our current encryption methods at risk; quantum key distribution will make encryption safe again; Grover's algorithm will speed up data searches. But what does all this really mean? How does it all work? All of this will be explained.

Can this be done without using mathematics? No, not if we want to really understand what is going on. The underlying ideas come from quantum mechanics and are often counterintuitive. Attempts to describe these in words don't work because we have no experience of them in our everyday lives. Even worse, verbal descriptions often give the impression that we have understood something when we really haven't. The good news is that we really do not need to introduce much mathematics. My role as a mathematician is to simplify the mathematics as much as possible—just sticking to the absolute essentials—and to give elementary examples to illustrate both how it is used and what it means. That said, the book probably contains mathematical ideas that you have not seen before, and, as with all mathematics, new concepts can seem strange at first. It is important not to gloss over the examples but to read them carefully, following each step of the calculations.

Quantum computing is a beautiful fusion of quantum physics with computer science. It incorporates some of the most stunning ideas of physics from the twentieth century into an entirely new way of thinking about computation. The basic unit of quantum computing is the qubit. We will see what qubits are and what happens when we measure them. A classical bit is either 0 or 1. If it's 0 and we measure it, we get 0. If it's 1 and we measure 1, we get 1. In both cases the bit remains unchanged. The situation is totally different for qubits. A qubit can be in one of an infinite number of states—a superposition of both 0 and 1—but when we measure it, as in the classical case, we just get one of two values, either 0 or 1. The act of measurement changes the qubit. A simple mathematical model describes all of this precisely.

Qubits can also be entangled. When we make a measurement of one of them, it affects the state of the other. Again, this is something that we don't experience in our daily lives, but it is described perfectly by our mathematical model.

These three things—superposition, measurement, and entanglement—are the key quantum mechanical ideas. Once we know what they mean, we can see how they may be used in computations. It is here that human ingenuity enters the picture.

Mathematicians often describe proofs as being beautiful, often containing unexpected insights. I feel exactly the same way about many of the topics we will look at. Bell's theorem, quantum teleportation, superdense coding, all are gems. The error correcting circuit and Grover's algorithm are absolutely amazing.

By the end of the book, you should understand the basic ideas that underlie quantum computing, and you will have seen some ingenious and beautiful constructions. You will also come to realize that quantum computing and classical computing are not two distinct disciplines, but that quantum computing is the more fundamental form of computing—anything that can be computed classically can be computed on a quantum computer. The qubit is the basic unit of computation, not the bit. Computation, in its essence, really means quantum computation.

Finally, it should be emphasized that this book is about the theory of quantum computation. It is about software, not hardware. We briefly mention hardware in places and occasionally talk about how to physically entangle qubits, but these topics are just asides. The book is not about how to build a quantum computer, but how to use one.

Here's a brief description of the book's contents.

Chapter 1. The basic unit of classical computing is the bit. Bits can be represented by anything that can be in one of two possible states. The standard example is an electrical switch that can be either on or off. The basic unit of quantum computing is the qubit. This can be represented by the spin of an electron or the polarization of a photon, but the properties of spin and polarization are not nearly as familiar to us as a switch being in the on or off position.

We look at the basic properties of spin, starting with Otto Stern and Walther Gerlach's classic experiment in which they studied the magnetic properties of silver atoms. We see what happens when we measure spin in a number of different directions. The act of making a measurement can affect the state of a qubit. There is also an underlying randomness associated with some of the measurements that we will need to explain.

The chapter concludes by showing that experiments analogous to those for spin can be performed using polarized filters and ordinary light.

Chapter 2. Quantum computing is based on an area of mathematics called linear algebra. Fortunately, we only need a few concepts. This chapter introduces and describes the linear algebra we need and illustrates how it is going to be used in the later chapters.

We introduce vectors and matrices and show how to calculate the length of vectors and how to tell whether or not two vectors are perpendicular. The chapter starts by just considering elementary operations on vectors and then shows how matrices give a simple way of doing a number of these calculations simultaneously.

It is not initially apparent that this material is going to be useful, but it is. Linear algebra forms the foundation of quantum computing. Since the rest of the book uses the mathematics introduced here, this chapter needs to be read carefully.

Chapter 3. This chapter shows how the previous two chapters are connected. The mathematical model of spin or, equivalently, that of polarization is given using linear algebra. This enables us to give the definition of a qubit and to describe exactly what happens when we measure it.

Several examples of measuring qubits in different directions are given. The chapter ends with an introduction to quantum cryptography, describing the BB84 protocol.

Chapter 4. This chapter describes what it means for two qubits to be entangled. Entanglement is difficult to describe in words, but it is easy to describe mathematically. The new mathematical idea is the tensor product. This is the simplest way of combining mathematical models of individual qubits to give one model that describes a collection of qubits.

Though the mathematics is straightforward, entanglement is not something that we experience in everyday life. When one of a pair of entangled qubits is measured, it affects the second qubit. This is what Albert Einstein, who disliked it, called "spooky action at a distance." We look at several examples.

The chapter concludes by showing that we can't use entanglement to communicate faster than the speed of light.

Chapter 5. We look at Einstein's concerns with entanglement and whether a hidden variable theory can preserve local realism. We go through the mathematics of Bell's inequality—a remarkable result that gives an experimental way of determining whether or not Einstein's argument is correct. As most people know, Einstein's view was wrong, but even Bell thought he would be proved correct.

Artur Ekert realized that the setup for the test of Bell's inequality could also be used both to generate a secure key to be used for cryptography and at the same time to test whether any eavesdroppers are present. We conclude the chapter with a description of this cryptographic protocol.

Chapter 6. The chapter starts with standard topics in computation: bits, gates, and logic. Then we briefly look at reversible computation and the ideas of Ed Fredkin. We show that both the Fredkin gate and the Toffoli gate are universal—you can build a complete computer using only Fredkin gates (or Toffoli gates). The chapter concludes with Fredkin's billiard-ball computer. This is not really needed for the rest of our story, but its sheer ingenuity demands that it be included.

This computer consists of balls colliding with one another and off various walls. It conjures up images of particles interacting. This is one of the ideas that inspired Richard Feynman to become interested in the idea of quantum computing. Feynman wrote some of the earliest papers on the subject.

Chapter 7. This chapter begins the study of quantum computing using quantum circuits. Quantum gates are defined. We see how a quantum gate acts on a qubit and realize that we have been considering these ideas all along. We just need to change our perspective. We no longer think of an orthogonal matrix as acting on our measuring device, but as acting on the qubit. We also prove some amazing results concerning superdense coding, quantum teleportation, cloning, and error correction.

Chapter 8. This is probably the most challenging chapter. In it we look at some quantum algorithms and show how quickly they can compute an answer compared to classical algorithms. To talk about the speed of algorithms we need to introduce various ideas from complexity theory. Once we have defined something called *query complexity*, we study three quantum algorithms and show that they are faster with respect to this type of complexity than their classical counterparts.

Quantum algorithms exploit the underlying structure of the problem that is being solved. It is much more than just the idea of quantum parallelism—putting the input into a superposition of all possible states. This chapter introduces the last piece of mathematical machinery, the Kronecker product of matrices. But the difficulty of the material is really caused by the fact that we are computing in a completely new way and we have no experience of thinking about solving problems using these novel ideas.

Chapter 9. The last chapter looks at the impact that quantum computing is going to have on our lives. We start by giving brief descriptions of two important algorithms, one invented Peter Shor, the other by Lov Grover.

Shor's algorithm gives a way of factoring a large number into the product of its prime factors. This might not seem that important, but our Internet security depends on this problem being hard to solve. Being able to factor products of large primes threatens our current methods of securing transactions between computers. It might be some time until we have quantum computers powerful enough to factor the large numbers that are currently in use, but the threat is real, and it is already forcing us to think about how to redesign the ways that computers can securely talk to one another.

Grover's algorithm is for special types of data searches. We show how it works for a small case and indicate how it works in general. Both Grover's and Shor's algorithms are important, not only for the problems they can solve but also for the new ideas they introduce. These underlying ideas have been and are being incorporated into a new generation of algorithms.

After looking at algorithms, we switch gears and briefly look at how quantum computation can be used to simulate quantum processes. Chemistry, at its most basic level, is quantum mechanical. Classical computational chemistry works by taking quantum mechanical equations and simulating them using classical computers. These simulations are approximations and ignore the fine details. This works well in many cases, but in some cases it doesn't. In these cases you need the fine details, and quantum computers should be able to give them.

This chapter also briefly looks at building actual machines. This is a very fast-growing area. The first machines are being offered for sale. There is even one machine available on the cloud that everyone can use for free. It looks likely that we will soon enter the age of *quantum supremacy*. (We explain what this means.)

The book concludes with the realization that quantum computation is not a new type of computation but is the discovery of the true nature of computation.

1 Spin

All computations involve inputting data, manipulating it according to certain rules, and then outputting the final answer. For classical computations, the *bit* is the basic unit of data. For quantum computations, this unit is the *quantum bit*—usually shortened to *qubit*.

A classical bit corresponds to one of two alternatives. Anything that can be in exactly one of two states can represent a bit. Later we will see various examples, which include the truth or falsity of a logical statement, a switch being in the on or off position, and even the presence or absence of a billiard ball.

A qubit, like a bit, includes these two alternatives, but—quite unlike a bit—it can also be in a combination of these two states. What does this mean? What exactly is a combination of two states, and what are physical objects that can represent qubits? What is the quantum computation analog to the switch?

A qubit can be represented by the spin of an electron or the polarization of a photon. This, though true, does not seem particularly helpful as spins of electrons and polarizations of photons are not things that most of us have knowledge about, let alone experience with. Let's start with a basic introduction to describe spin and polarization. To do this we describe the foundational experiment performed by Otto Stern and Walther Gerlach on the spin of silver atoms.

In 1922, Niels Bohr's planetary model described the current understanding of atoms. In this model an atom consisted of a positive nucleus orbited by negative electrons. These orbits were circular and were constrained to certain radii. The innermost orbit could contain at most two electrons. Once this was filled, electrons would start filling the next level, where at most eight electrons could be held. Silver atoms have 47 electrons. Two of these are in the innermost orbit, then eight in the next orbit, then eighteen more electrons in both the third and fourth levels. This leaves one lone electron in the outermost orbit.

Now, electrons moving in circular orbits generate magnetic fields. The electrons in the inner orbits are paired, each of the pair rotating in the opposite direction to its partner, resulting in their magnetic fields canceling. However, the single electron in the outer orbit generates a magnetic field that is not canceled by other electrons. This means that the atom as a whole can be considered as a little magnet with both a south pole and a north pole.

Stern and Gerlach designed an experiment to test whether the north–south axes of these magnets could have any direction whatsoever or whether they were constrained to certain directions. They did this by sending a beam of silver atoms through a pair of magnets as is depicted in figure 1.1. The vee-shaped design of the magnets makes the south magnet act more strongly than the north. If the silver atom is a magnet with north on top and south on bottom, it will be attracted to both the magnets of the apparatus, but the south magnet wins and the particle is deflected upward. Similarly if the silver atom is a magnet with south on top and north on bottom, it will be repelled by both the magnets of the apparatus, but again the

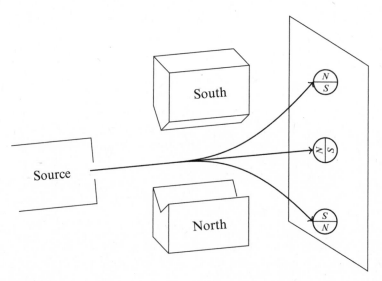

Figure 1.1
Stern-Gerlach apparatus.

south magnet wins and the particle is deflected downward. After passing through the apparatus, the atoms are collected on a screen.

From the classical viewpoint, the magnetic poles of the atom could be aligned in any direction. If they were aligned horizontally, there would be no deflection, and in general, the size of the deflection would correspond to the amount the magnetic axis of the atom differs from the horizontal, with maximum deflections occurring when the magnetic poles of the atom are aligned vertically.

If the classical viewpoint is correct, when we send a large number of silver atoms through the machine we ought to see a continuous line on the screen going from the top point to the bottom. But this is not what Stern and Gerlach found. When they looked at the screen, they found just two dots: one at the extreme top and the other at the extreme bottom. All of the atoms behaved like little bar magnets that were aligned vertically. None of them had any other orientation. How could this be?

But before we start analyzing what is going on in more detail, we will shift our attention from atoms to electrons. Not only do atoms act like little magnets, but so also do their components. When we discuss quantum computers we will often talk about electrons and their spins. As with silver atoms, if you measure spin* in the vertical direction, you find that the electron is either deflected in the north direction or the south direction. Again, like silver atoms, you find that electrons are little magnets with their north and south poles perfectly aligned in the vertical direction. None of them have any other orientation.

In practice, you can't actually measure electron spin of a free electron using the Stern-Gerlach apparatus in the way we have shown because electrons have a negative charge and magnetic fields deflect moving charged particles. That said, the following diagrams give useful pictorial representations of the results of measuring spin in various directions. The idea behind this diagram is that you are the source; the magnets are lined up between you and this book. The dot shows how the electron gets deflected. In figure 1.2, the picture on the left shows the deflection by the magnets. The one on the right gives a depiction of the electron as a magnet with the north and south poles marked. We will describe this situation as saying the electron

---

* We will keep using the term *spin* because it is the standard terminology. But we are just determining the axis of the poles of a magnet.

(a) Outcome of experiment   (b) Diagram of electron

Figure 1.2
Electron with spin N in the vertical direction.

(a) Outcome of experiment   (b) Diagram of electron

Figure 1.3
Electron with spin S in the vertical direction.

has *spin N in the vertical direction*. Figure 1.3 shows the other possibility, where the electron has *spin S in the vertical direction*.

To understand the deflection, it helps to remember that the south magnet acts more strongly than the north, and so to calculate the direction of deflection you just consider the effect of this magnet. If the electron is aligned with its north pole closest to the south magnet, then it will be attracted and the deflection will be in the direction of the south magnet. If the electron is aligned with its south pole closest to the south magnet, then it will be repelled, and the deflection will be in the direction of the north magnet.

Of course, there is nothing special about the vertical direction. For example, we can rotate the magnets through 90°. The electrons will still be deflected in the direction given by either the north magnet or the south magnet. In this case, the electrons now behave as magnets with their north and south poles aligned in the horizontal direction, as is depicted in figures 1.4 and 1.5.

(a) Outcome of experiment   (b) Diagram of electron

Figure 1.4
Electron with spin N in the 90° direction.

(a) Outcome of experiment   (b) Diagram of electron

Figure 1.5
Electron with spin S in the 90° direction.

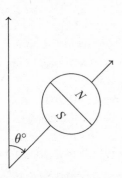

Figure 1.6
Electron with spin N in the $\theta°$ direction.

In the following chapters we will want to rotate the magnets through various angles. We will measure angles in the clockwise direction with 0° denoting the upward vertical direction and θ measuring the angle from the upward vertical. Figure 1.6 depicts an electron with spin N in the direction of a general angle $\theta°$.

Sometimes spin is described as being *up, down, left,* or *right.* Our description of an electron being N in the direction 0° seems somewhat cumbersome, but it is unambiguous and avoids some of the pitfalls of using *up, down,* and so on, especially when we rotate the apparatus through 180°. For example, both of the situations pictured in figure 1.7 represent an electron having spin N in direction 0° or equivalently spin S in direction 180°.

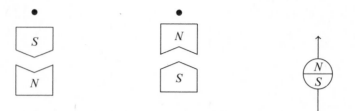

(a) Outcome of experiment   (b) Outcome of experiment   (c) Diagram of electron

Figure 1.7
Electron with spin N in the 0° direction.

Before we proceed with our study of electron spin, we will pause and look at an analogy that we will use in several places.

The Quantum Clock

Imagine that you have a clock with a dial marked with hours in the standard positions. It also has a hand. You are, however, forbidden to look at the face of the clock. You can only ask it questions. You want to know in which direction the hand is pointing, so you would like to ask the clock this seemingly simple question. But it is not allowed. You are only allowed to ask whether the hand is pointing at a particular number on the face. So, for example, you can ask if the hand is pointing to twelve, or you can ask if it is pointing to four. Now, if this were a regular clock you would have to be extremely lucky to get a yes answer. Most of the time the hand would be pointing in a completely different direction. But the quantum clock is not like a regular clock. It either answers *yes* or it tells you that the hand is pointing in the direction exactly opposite the one you asked about. If we ask if the hand is pointing in the direction of twelve, it will tell us either that it is or that it is pointing in the direction of six. If we ask if the hand is pointing in the direction of four, it will either tell us it is, or that it is pointing in the direction of ten. This is a very curious state of affairs, but it is exactly analogous to electron spin.

As we said, electron spin is going to be the idea that motivates the definition of the qubit. If we are going to do computations, we need to understand the rules that govern spin measurements. We start by considering what happens when we measure more than once.

Measurements in the Same Direction

Measurements are repeatable. If we repeat exactly the same measurement, we get exactly the same result. For example, suppose that we decide to measure an electron's spin in the vertical direction. We then repeat exactly the same experiment by positioning two more sets of our apparatus behind the first one. One is positioned in exactly the right place to catch electrons that are deflected upward by the first apparatus. The other is placed to catch the electrons deflected downward. The electrons that are deflected upward by the first apparatus are deflected up by the second, and the ones deflected down by the first apparatus are deflected down by the second. This means that electrons measured to have spin N in direction $0°$ initially also have spin N in direction $0°$ when we repeat the experiment. Similarly, if an electron is initially measured to have spin S in direction $0°$ and we repeat exactly the same experiment, it will still have spin S in direction $0°$. For our clock analogy, if we repeatedly ask if the hand is pointing at twelve, we will repeatedly get the same answer: that either it is always pointing toward twelve or it is always pointing toward six.

There is, of course, nothing special about the vertical direction. If we start by measuring in direction $\theta°$, and then repeatedly measure in the same direction, we will obtain the exactly the same result each time. We will end up with a string of letters consisting entirely of Ns or one entirely of Ss.

The next thing to consider is what happens if we don't repeat the same measurement. For example, what happens if we first measure vertically and then horizontally?

Measurements in Different Directions

We will measure the electron's spin first in the vertical direction, then in the horizontal direction. We will send a stream of electrons through the first detector—measuring spin in the vertical direction. As before, we have two more detectors behind the first one in the appropriate positions to catch the electrons coming from the first detector. The difference is that these two detectors are rotated through $90°$ and measure spin in the horizontal direction.

First we look at the stream of electrons that are deflected upward by the first detector—these have spin N in direction $0°$. When they go through the second detector, we find that half of them have spin N and half have spin S in direction 90°. The sequence of north and south spins in direction 90° is completely random. There is no way of telling whether an electron that had spin N in direction 0° will have either spin S or N when we measure it again in direction 90°. Similar results hold for the electrons that the first detector shows have spin S in the vertical direction—exactly half have spin N in the horizontal direction, and the other half have spin S in the horizontal direction. Again, the sequence of Ns and Ss is completely random.

The analogous questions for our clock are asking about whether the hand is pointing in the direction of twelve and then asking if it is pointing in the direction of three. If we have a large number of clocks and ask them these two questions, the answers to the second questions will be random. Half of the clocks will say the hand is pointing in the direction of three. The other half will say in the direction of nine. The answers to the first question have no bearing on the answers to the second question.

Finally, we will look at what happens when we make three measurements. First we measure vertically, then horizontally, and then vertically once more. Consider a stream of electrons coming from the first detector that have spin N in direction 0°. We know that half of them will have spin N and half have spin S when we measure spin in direction 90°. We will restrict attention to the stream that corresponds to N for the first two measurements and then, for the third measurement, measure spin in the vertical direction. We find that exactly half of these electrons have spin N in direction 0° and half have spin S. Once more the sequence of Ns and Ss is completely random. The fact that the electrons initially had spin N in the vertical direction has no bearing on whether or not they have will still have spin N when we again measure in the vertical direction.

What conclusions can we draw from these results? There are three. And they are all important.

First, if we keep repeating exactly the same question we get exactly the same answer. This tells us that sometimes there are definite answers. We are not getting random answers to every question.

Second, randomness does seem to occur. If we ask a sequence of questions, the final results can be random.

Third, measurements affect outcomes. We saw that if we ask the same question three times, we get the exactly the same answer three times. But if the first and third questions are identical and the second is different, the answers to the first and third questions need not be the same. For example, if we ask three times in a row if the hand is pointing toward twelve, we will get exactly the same answer each time, but if we ask first if it is pointing toward twelve, then whether it is pointing to three, and finally again whether it is pointing toward twelve, the answers to the first and third question need not be the same. The only difference between the two scenarios is the second question, so that question must be affecting the outcome of the following question. We will say a little more about these observations, starting with measurements.

Measurements

In classical mechanics, we might consider the path of a ball thrown into the air. The path can be calculated using calculus, but in order to perform the calculation we need to know certain quantities such as the mass of the ball and its initial velocity. How we measure these is not part of the theory. We just assume that they are known. The implicit assumption is that the act of measuring is not important to the problem—that taking a measurement does not affect the system being modeled. For the example of a ball being thrown into the air, this makes sense. We can measure its initial velocity using a radar gun, for example. This involves bouncing photons off the ball and, though bouncing photons will have an effect on the ball, it is negligible. This is the philosophy underlying classical mechanics: Measurements will affect the objects being studied, but experiments can be designed so the effect of measurement is negligible and so can consequently be ignored.

In quantum mechanics, we are often considering tiny particles like atoms or electrons. Here bouncing photons off them has an effect that is no longer negligible. In order to perform some measurement, we have to interact with the system. These interactions are going to perturb our system, so we can no longer ignore them. It should not seem surprising that measurement becomes a basic component of the theory, but what is surprising is how this is done. For example, consider the case where we measure the spin of an electron first in the vertical direction and then in the horizontal one. We have seen that exactly half of the electrons that have spin N in direction $0°$ after passing through the first detector will have spin N in direction $90°$ when measured by the second detector. It might seem that the strength of the magnets might be having some effect on the outcome, perhaps they are so strong that they are causing the magnetic axes of the electrons to twist to align with the magnetic field of the measuring device, and that if we had weaker magnets the twisting would be lessened and we might get a different result. However, this is not how measurement is incorporated into the theory. As we shall see, our model does not take into account the "strength" of the measurement. Rather, it is the actual process of taking the measurement, however it is done, that has an effect on the system. Later we will describe the mathematics that models how measuring spin is treated in quantum mechanics. Each time a measurement is made, we will see that the system is changed in certain prescribed ways; these prescribed ways depend on the type of measurement being made but not on the strength of the measurement.

Incorporating measurements into the theory is one on the differences between classical and quantum mechanics. Another difference concerns randomness.

Randomness

Quantum mechanics involves randomness. For example, if we first measure the spin of a stream of electrons in the vertical direction, then in the horizontal direction, and record the results from the second measuring device, we will obtain a string of Ns and Ss. This sequence of spins is completely random. For example, it might look something like *NSSNNNSS.* ...

The classical experiment for generating a random sequence of two symbols each associated with probability of a half is that of tossing a fair coin. If we toss a fair coin we might get a sequence *HTTHHHTT.* ... Although these two examples yield similar results, there is a big difference in how randomness is interpreted in the two theories.

Tossing a coin is something that is described by classical mechanics. It can be modeled using calculus. To compute whether the coins lands heads or tails up, you need first to carefully measure the initial conditions: the weight of the coin, the height above the ground, the force of the impact of the thumb on the coin, the exact location on the coin where the thumb hits, the position of the coin, and so forth and so on. Given all of these values exactly, the theory will tell us which way up the coin lands. There is no actual randomness involved. Tossing a coin seems random because

each time we do it the initial conditions vary slightly. These slight variations can change the outcome from heads to tails and vice versa. There is
no real randomness in classical mechanics, just what is often called *sensitive
dependence to initial conditions*—a small change in the input can get amplified and produce an entirely different outcome. The underlying idea concerning randomness in quantum mechanics is different. The randomness
is true randomness.

The sequence *NSSNNNSS* ... that we obtained from measuring spin in two
directions is considered to be truly random, as we shall see. The sequence
of coin tosses, *HTTHHHTT* ... appears random, but the classical laws of
physics are deterministic and this apparent randomness would disappear if
we could make our measurements with infinite accuracy.

At this stage it is natural to question this. Einstein certainly did not like
this interpretation, famously saying that God does not play dice. Couldn't
there be a deeper theory? If we knew more information about the initial
configurations of our electrons, couldn't it be the case that the final results
would no longer be random but completely determined? Couldn't there be
hidden variables—once we know the values of these variables, the apparent
randomness disappears? In what follows we will present the mathematical theory in which true randomness is used. Later we will return to these
questions. We will describe a clever experiment to distinguish between the
hidden variable and the true randomness hypotheses. This experiment has
been performed several times. The outcomes have always shown that the
randomness is real and that there is no simple hidden variable theory that
can eliminate it.

We started this chapter by saying that a qubit can be represented by the
spin of an electron or the polarization of a photon. We will show how the
models for spin and polarization are related.

Photons and Polarization

It is often said that we are not aware of the strange quantum phenomena because they only occur at incredibly small scales and are not apparent at the scales of our everyday life. There is some truth to this, but
there is an experiment that is completely analogous to measuring spin
of electrons that can be performed with very little apparatus. It concerns
polarized light.

To perform these experiments you need three squares of linear polarized film. Start by taking two of the squares and putting one in front of the other. Keep one square fixed and rotate the other by ninety degrees. You will find that light passes through the pair of filters when they are aligned in one direction, but is completely blocked when one of the filters is rotated by ninety degrees. This is not particularly exciting. But now rotate the two filters so that no light passes through, take the third filter, rotate it by forty-five degrees, and slide it between the other two. Amazingly, light passes through the region where the three filters overlap—no light passes through the overlap of just the original two filters, but it does where all three overlap.

I heard about this experiment with three filters several years ago. I asked a friend who is a physicist if he had any polarized sheet. He invited me to his lab, where he had an enormous roll of it. He cut a piece off and gave it to me. I used scissors to cut it into three squares of about an inch by an inch and performed the experiment—and it worked! This experiment is so simple and yet so surprising. I have kept the three squares in my wallet ever since.

When we measure polarization we find that photons are polarized in two perpendicular directions, both of which are perpendicular to the direction of travel of the photon. The polarized square lets through photons that are polarized in one of the two directions and absorb the photons that are polarized in the other. The polarized squares correspond to the Stern-Gerlach apparatus. Sending light through a square can be considered making a measurement. As with spin, there are two possible outcomes: Either the direction of polarization is directly aligned with the orientation of the square, in which case the photon passes through, or the direction of polarization is perpendicular to the orientation of the square, in which case the photon is absorbed.

We start by assuming that our square has vertical orientation so that it lets through photons with vertical polarization and absorbs the ones with horizontal polarization, and consider a number of experiments that correspond to the ones we described for electron spin.

First, suppose that we have two squares, both with the same orientation, so they both let through photons with vertical polarization. If we look at the squares individually they look gray, as is expected. They are both absorbing some photons—those with horizontal polarization. If we then slide one of the squares over the other, there is minimal change. The amount of light let through the two overlapping squares is about the same as the amount that comes through each square when they are not overlapping. This is depicted in figure 1.8.

We will now rotate one of the squares through ninety degrees. Assuming we are not looking at light reflecting off a shiny surface, or light coming directly from a computer screen, but we are in normal light conditions, the proportion of horizontally polarized photons is equal to the proportion of vertically polarized ones, and both squares will look equally gray. We repeat the experiment of overlapping these squares. This time no light is let through the region of overlap, as depicted in figure 1.9.

The third experiment is to take the third sheet and rotate it through forty-five degrees. Under normal light conditions nothing appears to happen as we rotate the square. It maintains the same shade of gray. We now slide this square between the other two squares, one of which has vertical orientation, and the other has horizontal orientation. The result, as we noted earlier, is both surprising and unintuitive. Some light comes through the region of overlap of all three squares. (This is depicted in figure 1.10.) These polarized squares are sometimes called filters, but clearly they are

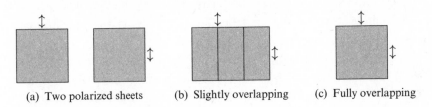

(a) Two polarized sheets　　(b) Slightly overlapping　　(c) Fully overlapping

Figure 1.8
Two linear polarized squares with the same orientation.

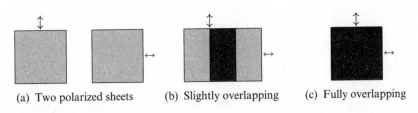

(a) Two polarized sheets　　(b) Slightly overlapping　　(c) Fully overlapping

Figure 1.9
Two linear polarized squares with different orientations.

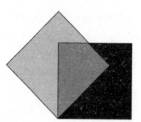

Figure 1.10
Three linear polarized squares with different orientations.

not acting in the conventional ways that filters work. More light comes through three filters than comes through two!

We will give a brief description of what is happening. Later we will see the mathematical model that describes both spin and polarization.

Recall our quantum clock. We can ask if the hand is pointing at twelve, or we can ask if the hand is pointing at six. The information we gain from either question tells us which of the numbers 12 or 6 the hand is pointing to, but the Yes/No answers are reversed. For the polarized squares the analogous questions are asked by rotating the square by ninety degrees—not one hundred and eighty. The information we obtain is the same. The difference is that if the answer is yes, the photon passes through the filter and we can perform more measurements on it, but if the answer is no, the filter absorbs the photon, so we cannot ask it further questions.

The first two experiments involved just two sheets and are telling us exactly the same thing: When we repeat a measurement, we get the same result. In both experiments we are measuring the polarization in the vertical and horizontal directions two times. In these experiments, the photons that pass through the first filter have vertical orientations. The first experiment, where the second filter also has vertical orientation, we are asking the question, "Is the photon vertically polarized?" twice and we receive the answer "Yes" twice. In the second experiment, the second question is changed to "Is the photon horizontally polarized?" and receives the answer "No." Both experiments give us the same information, but the negative answer for the second question in the second experiment means that the photon is absorbed and so, unlike the first experiment, it is not available for further questioning.

In the third experiment, the filter that has been rotated through forty-five degrees is now measuring the polarization at angles of 45° and 135°. We know that the photons coming through the first filter are polarized vertically. When measured by the second filter, half of the photons are found to be polarized in the 45° direction and half in the 135° directions. The ones with 45° polarization pass through the filter, and the others are absorbed. The third filter again measures the polarization in the vertical and horizontal directions. The photons entering have 45° polarization, and when measured in the vertical and horizontal directions, half will have vertical polarization and half will have horizontal polarization. The filter absorbs the vertically polarized photons and lets through those that are polarized horizontally.

Conclusions

We started this chapter by saying that classical bits can be represented by everyday objects like switches in the on or off position, but that qubits are generally represented by the spin of electrons or the polarization of photons. Spin and polarization are not nearly so familiar to us and have properties that are quite unlike their classical counterparts.

To measure spin, you first have to choose a direction and then measure it in that direction. Spin is *quantized*: When measured, it gives just two possible answers—not a continuous range of answers. We can assign classical bits to these results. For example, if we obtain an *N* we can consider it to be the binary digit 0, and if we obtain an *S* we can consider it to be the binary digit 1. This is exactly how we get answers from a quantum computation. The last stage of the computation is to take a measurement. The result will be one of two things, which will be interpreted as either 0 or 1. Although the actual computation will involve qubits, the final answer will be in terms of classical bits.

We have only just started our study, so we are quite limited in what we can do. We can, however, generate random strings of binary digits. The experiment that generated random strings of *N*s and *S*s can be rewritten as a string of 0s and 1s. Consequently measuring spins of electrons first in the vertical and then in the horizontal direction gives a random string of 0s and 1s. This is probably the simplest thing that we can do with qubits, but surprisingly this is something that cannot be done with a classical computer. Classical computers are deterministic. They can compute strings that pass various tests for randomness, but these are pseudorandom, not random. They are computed by some deterministic function, and if you know the function and the initial seed input, you can calculate exactly the same string. There are no classical computer algorithms that generate truly random strings. Thus, already we can see that quantum computations have some advantages over classical ones.

Before we start to describe other quantum computations we need to develop a precise mathematical model that describes what happens when we measure spin in various directions. This is started in the next chapter where we study linear algebra—the study of the algebra associated with vectors.

2 Linear Algebra

Quantum mechanics is based on linear algebra. The general theory uses infinite dimensional vector spaces. Fortunately for us, to describe spin or polarization we need only finite dimensions, which makes things much easier. In fact, we need only a few tools. At the end of this chapter I have given a list. The rest of the chapter explains how to use these tools and what the calculations mean. There are many examples. It is important to work carefully through all of them. The mathematics introduced here is essential to everything that follows. Like much mathematics, it can seem complicated when it is first introduced, but it becomes almost second nature with practice. The actual computations only involve addition and multiplication of numbers, along with an occasional square root and trigonometric function.

We will be using Paul Dirac's notation. Dirac was one of the founders of quantum mechanics, and his notation is used extensively throughout both quantum mechanics and quantum computing. It is not widely used outside these disciplines, which is surprising given how elegant and useful it is.

But first, we begin with a brief description of the numbers we will be using. These are real numbers—the standard decimal numbers with which we are all familiar. Practically every other book on quantum computation uses complex numbers—these involve the square root of negative one. So, let's start by explaining why we are not going to be using them.

Complex Numbers versus Real Numbers

Real numbers are straightforward to use. Complex numbers are—well, more complicated. To talk about these numbers we would have to talk about their moduli and explain why we have to take conjugates. For what we are going to do, complex numbers are not needed and would only add another layer of difficulty. Why then, you ask, do all the other books use complex numbers? What can you do with complex numbers that you cannot do with real ones? Let's briefly address these questions.

Recall that we measured the spin of an electron at various angles. These angles are all in one plane, but we live in a three-dimensional world. We compared measuring spin to using our quantum clock. We could only ask about directions given by the hand moving around the two-dimensional face. If we move to three dimensions, our analog would not be a clock face, but a globe with the hand at its center pointing to locations on the surface. We could ask, for example, if the hand is pointing to New York. The answer would be either that it is, or that it is pointing to the point diametrically opposite New York. The mathematical model for spin in three dimensions uses complex numbers. The computations involving qubits that we will look at, however, need to measure spin in only two dimensions. So, though our description using real numbers is not quite as encompassing as that using complex numbers, it is all that we need.

Finally, complex numbers provide an elegant way of connecting trigonometric and exponential functions. At the very end of the book we will look at Shor's algorithm. This would be hard to explain without using complex numbers. But this algorithm also needs continued fractions, along with results from number theory and results about the speed of an algorithm for determining whether a number is prime. There would be a significant jump in the level of mathematical sophistication and knowledge needed if we were to describe Shor's algorithm in full detail. Instead we will describe the basic ideas that underlie the algorithm, indicating how these fit together. Once again, our description will use only real numbers.

So, for what we are going to do, complex numbers are not needed. If, however, after reading this book, you want to continue studying quantum computation, they will be needed for more advanced topics.

Now that we have explained why we are going to stay with the real numbers, we begin our study of vectors and matrices.

Vectors

A *vector* is just a list of numbers. The *dimension* of the vector is the number of numbers in the list. If the lists are written vertically, we call them *column vectors* or *kets*. If the lists are written horizontally, we call them *row vectors* or *bras*. The numbers that make up a vector are often called *entries*. To illustrate, here is a three-dimensional ket and a four-dimensional bra:

$$\begin{bmatrix} 2 \\ 0.5 \\ -3 \end{bmatrix}, [1 \quad 0 \quad -\pi \quad 23].$$

The names *bra* and *ket* come from Paul Dirac. He also introduced notation for naming these two types of vectors: a ket with name v is denoted by $|v\rangle$; a bra with name w is denoted by $\langle w|$. So we might write

$$|v\rangle = \begin{bmatrix} 2 \\ 0.5 \\ -3 \end{bmatrix} \text{ and } \langle w| = [1 \quad 0 \quad -\pi \quad 23].$$

Later we will see why we use two different symbols to surround the name, and the reason that tells us which side the angled bracket goes. But, for now, the important thing is to remember that kets refer to columns (think of the repeated "k" sound) and that bras, as usual, have their entries arranged horizontally.

Diagrams of Vectors

Vectors in two or three dimensions can be pictured as arrows. We will look at an example using $|a\rangle = \begin{bmatrix} 3 \\ 1 \end{bmatrix}$. (In what follows we will often use kets for our examples, but if you like you can replace them with bras.) The first entry, 3 in this example, gives the change in the x-coordinate from the initial point to the terminal point. The second entry gives the change in the y-coordinate going from the initial point to terminal point. We can draw this vector with any initial point—if we choose (a, b) as the coordinates of its initial point, then the coordinates of its terminal point will be at $(a+3, b+1)$. Notice that if the initial point is drawn at the origin, the terminal point has coordinates given by the entries of the vector. This is convenient,

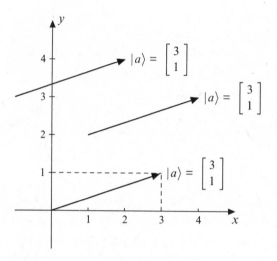

Figure 2.1
Same ket drawn in different positions.

and we will often draw them in this position. Figure 2.1 shows the same ket drawn with different initial points.

Lengths of Vectors

The length of a vector is, as might be expected, the distance from its initial point to its terminal point. This is the square root of the sum of squares of the entries. (This comes from the Pythagorean theorem.) We denote the length of a ket $|a\rangle$ by $\||a\rangle|$, so for $|a\rangle = \begin{bmatrix} 3 \\ 1 \end{bmatrix}$ we have $\||a\rangle| = \sqrt{3^2 + 1^2} = \sqrt{10}$.

More generally, if $|a\rangle = \begin{bmatrix} a_1 \\ a_2 \\ \vdots \\ a_n \end{bmatrix}$, then $\||a\rangle| = \sqrt{a_1^2 + a_2^2 + \cdots + a_n^2}$.

Vectors of length 1 are called *unit* vectors. Later we will see that qubits are represented by unit vectors.

Scalar Multiplication

We can multiply a vector by a number. (In linear algebra, numbers are often called scalars. Scalar multiplication just refers to multiplying by a number.) We do this by multiplying each of the entries by the given number. For example, multiplying the ket $|a\rangle = \begin{bmatrix} a_1 \\ a_2 \\ \vdots \\ a_n \end{bmatrix}$ by the number c gives $c|a\rangle = \begin{bmatrix} ca_1 \\ ca_2 \\ \vdots \\ ca_n \end{bmatrix}$.

It is straightforward to check that multiplying a vector by a positive number c multiplies its length by a factor of c. We can use this fact to enable us to get vectors of different lengths pointing in the same direction. In particular, we will often want to have a unit vector pointing in the direction given by a non–unit vector. Given any non-zero vector $|a\rangle$, its length is $\|a\rangle\|$. If we multiply $|a\rangle$ by the reciprocal of its length, we obtain a unit vector. For example, as we have already seen, if $|a\rangle = \begin{bmatrix} 3 \\ 1 \end{bmatrix}$ then $\|a\rangle\| = \sqrt{10}$.

If we let

$$|u\rangle = \frac{1}{\sqrt{10}}\begin{bmatrix} 3 \\ 1 \end{bmatrix} = \begin{bmatrix} \dfrac{3}{\sqrt{10}} \\ \dfrac{1}{\sqrt{10}} \end{bmatrix},$$

then

$$\|u\rangle\| = \sqrt{\left(\frac{3}{\sqrt{10}}\right)^2 + \left(\frac{1}{\sqrt{10}}\right)^2} = \sqrt{\frac{9}{10} + \frac{1}{10}} = \sqrt{1} = 1.$$

Consequently, $|u\rangle$ is a unit vector that points in the same direction as $|a\rangle$.

Vector Addition

Given two vectors that have the same type—they are both bras or both kets—and they have the same dimension, we can add them to get a new vector of the same type and dimension. The first entry of this vector just comes from adding the first entries of the two vectors, the second entry from adding the two second entries, and so on. For example, if $|a\rangle = \begin{bmatrix} a_1 \\ a_2 \\ \vdots \\ a_n \end{bmatrix}$

and $|b\rangle = \begin{bmatrix} b_1 \\ b_2 \\ \vdots \\ b_n \end{bmatrix}$, then $|a+b\rangle = \begin{bmatrix} a_1 + b_1 \\ a_2 + b_2 \\ \vdots \\ a_n + b_n \end{bmatrix}$.

Vector addition can be pictured by what is often called the *parallelogram law for vector addition*. If the vector $|b\rangle$ is drawn so that its initial point is at the terminal point of $|a\rangle$, then the vector that goes from the initial point of $|a\rangle$ to the terminal point of $|b\rangle$ is $|a+b\rangle$. This can be drawn giving a triangle.

We can interchange the roles of $|a\rangle$ and $|b\rangle$, drawing the initial point of $|a\rangle$ at the terminal point of $|b\rangle$. The vector that goes from the initial point of $|b\rangle$ to the terminal point of $|a\rangle$ is $|b+a\rangle$. Again, this gives a triangle. But we know that $|a+b\rangle = |b+a\rangle$. So if we draw the triangle construction for $|a+b\rangle$ and $|b+a\rangle$ where both the vectors have the same initial and terminal points, the two triangles connect to give us a parallelogram with the diagonal representing both $|a+b\rangle$ and $|b+a\rangle$. Figure 2.2 illustrates this where $|a\rangle = \begin{bmatrix} 3 \\ 1 \end{bmatrix}$, $|b\rangle = \begin{bmatrix} 1 \\ 2 \end{bmatrix}$, and consequently $|a+b\rangle = |b+a\rangle = \begin{bmatrix} 4 \\ 3 \end{bmatrix}$.

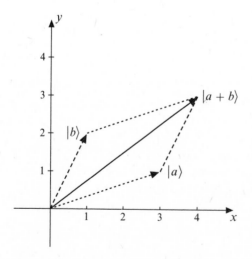

Figure 2.2
Parallelogram law for vector addition.

Orthogonal Vectors

Figure 2.2 helps us visualize some basic properties of vector addition. One of the most important comes from the Pythagorean theorem. We know that if a, b, and c represent the lengths of the three sides of a triangle, then $a^2 + b^2 = c^2$ if and only if the triangle is a right triangle. The picture then tells us that two vectors $|a\rangle$ and $|b\rangle$ are perpendicular if and only if $\||a\rangle|^2 + \||b\rangle|^2 = \||a+b\rangle|^2$.

The word *orthogonal* means exactly the same thing as perpendicular, and it is the word that is usually used in linear algebra. We can restate our observation: Two vectors $|a\rangle$ and $|b\rangle$ are orthogonal if and only if $\||a\rangle|^2 + \||b\rangle|^2 = \||a+b\rangle|^2$.

Multiplying a Bra by a Ket

If we have a bra and a ket of the same dimension, we can multiply them—the bra on the left and the ket on the right—to obtain a number. This is done in the following way, where we suppose that both $\langle a|$ and $|b\rangle$ are n-dimensional:

$$\langle a| = [a_1 \quad a_2 \quad \cdots \quad a_n] \quad \text{and} \quad |b\rangle = \begin{bmatrix} b_1 \\ b_2 \\ \vdots \\ b_n \end{bmatrix}.$$

We use concatenation to denote the product. This just means that we write down the terms side by side with no symbol between them. So the product is written $\langle a||b\rangle$. By squeezing the symbols even closer the vertical lines coincide and we get $\langle a|b\rangle$, which is the notation we will use. The definition of the bra-ket product is

$$\langle a|b\rangle = [a_1 \quad a_2 \quad \cdots \quad a_n] \begin{bmatrix} b_1 \\ b_2 \\ \vdots \\ b_n \end{bmatrix} = a_1 b_1 + a_2 b_2 + \cdots + a_n b_n.$$

The vertical lines of the bras and kets are "pushed together," which helps us to remember that the bra has the vertical line on the right side and the ket has it on the left. The result consists of terms sandwiched between angle brackets. The names "bra" and "ket" come from "bracket," which is almost the concatenation of the two names. Though this is a rather weak play on words, it does help us to remember that, for this product, that the "bra" is to the left of the "ket."

In linear algebra this product is often called the *inner* product or the *dot* product, but the bra-ket notation is the one used in quantum mechanics, and it is the one that we will use throughout the book.

Now that we have defined the bra-ket product, let's see what we can do with it. We start by revisiting lengths.

Bra-kets and Lengths

If we have a ket denoted by $|a\rangle$, then the bra $\langle a|$ with the same name is defined in the obvious way. They both have exactly the same entries, but for $|a\rangle$ they are arranged vertically, and for $\langle a|$ horizontally.

$$|a\rangle = \begin{bmatrix} a_1 \\ a_2 \\ \vdots \\ a_n \end{bmatrix} \quad \langle a| = [a_1 \quad a_2 \quad \cdots \quad a_n].$$

Consequently, $\langle a|a\rangle = a_1^2 + a_2^2 + \ldots + a_n^2$, and so the length of $|a\rangle$ can be written succinctly as $\||a\rangle\| = \sqrt{\langle a|a\rangle}$.

To illustrate, we return to the example where we found the length of $|a\rangle = \begin{bmatrix} 3 \\ 1 \end{bmatrix}$: $\langle a|a\rangle = [3 \quad 1]\begin{bmatrix} 3 \\ 1 \end{bmatrix} = 3^2 + 1^2 = 10$. Then we take the square root to obtain $\||a\rangle\| = \sqrt{10}$.

Unit vectors are going to become very important in our study. To see whether a vector is unit—has length 1—we will repeatedly use the fact that a ket $|a\rangle$ is a unit vector if and only if $\langle a|a\rangle = 1$.

Another important concept is orthogonality. The bra-ket product can also tell us when two vectors are orthogonal.

Bra-kets and Orthogonality

The key result is: Two kets $|a\rangle$ and $|b\rangle$ are orthogonal if and only if $\langle a|b\rangle = 0$. We will look at a couple of examples and then give an explanation of why this result is true.

Let $|a\rangle = \begin{bmatrix} 3 \\ 1 \end{bmatrix}$, $|b\rangle = \begin{bmatrix} 1 \\ 2 \end{bmatrix}$ and $|c\rangle = \begin{bmatrix} -2 \\ 6 \end{bmatrix}$. We calculate $\langle a|b\rangle$ and $\langle a|c\rangle$.

$$\langle a|b\rangle = \begin{bmatrix} 3 & 1 \end{bmatrix} \begin{bmatrix} 1 \\ 2 \end{bmatrix} = 3 + 2 = 5$$

$$\langle a|c\rangle = \begin{bmatrix} 3 & 1 \end{bmatrix} \begin{bmatrix} -2 \\ 6 \end{bmatrix} = -6 + 6 = 0$$

Since $\langle a|b\rangle \neq 0$, we know that $|a\rangle$ and $|b\rangle$ are not orthogonal. Since $\langle a|c\rangle = 0$, we know that $|a\rangle$ and $|c\rangle$ are orthogonal.

Why does this work? Here is an explanation for two-dimensional kets.

Let $|a\rangle = \begin{bmatrix} a_1 \\ a_2 \end{bmatrix}$ and $|b\rangle = \begin{bmatrix} b_1 \\ b_2 \end{bmatrix}$, then $|a\rangle + |b\rangle = \begin{bmatrix} a_1 + b_1 \\ a_2 + b_2 \end{bmatrix}$. We calculate the square of the length of $|a\rangle + |b\rangle$.

$$\||a\rangle + |b\rangle\|^2 = \begin{bmatrix} a_1 + b_1 & a_2 + b_2 \end{bmatrix} \begin{bmatrix} a_1 + b_1 \\ a_2 + b_2 \end{bmatrix}$$
$$= (a_1 + b_1)^2 + (a_2 + b_2)^2$$
$$= (a_1^2 + 2a_1b_1 + b_1^2) + (a_2^2 + 2a_2b_2 + b_2^2)$$
$$= (a_1^2 + a_2^2) + (b_1^2 + b_2^2) + 2(a_1b_1 + a_2b_2)$$
$$= \||a\rangle\|^2 + \||b\rangle\|^2 + 2\langle a|b\rangle$$

Clearly this number equals $\||a\rangle\|^2 + \||b\rangle\|^2$ if and only if $2\langle a|b\rangle = 0$. Now recall our observation that two vectors $|a\rangle$ and $|b\rangle$ are orthogonal if and only if $\||a\rangle\|^2 + \||b\rangle\|^2 = \||a + b\rangle\|^2$. We can restate this observation using our calculation for the square of the length of $|a\rangle + |b\rangle$ to say: Two vectors $|a\rangle$ and $|b\rangle$ are orthogonal if and only if $\langle a|b\rangle = 0$.

Though we have shown this for two-dimensional kets, the same argument can be extended to kets of any size.

Orthonormal Bases

The word "orthonormal" has two parts; ortho from orthogonal, and normal from normalized which, in this instance, means unit. If we are working with two-dimensional kets, an orthonormal basis will consist of a set of two unit kets that are orthogonal to one another. In general, if we are working with n-dimensional kets, an orthonormal basis consists of a set of n unit kets that are mutually orthogonal to one another.

We begin by looking at two-dimensional kets. The set of all two-dimensional vectors is denoted by $\mathbb{R}^2$. An orthonormal basis for $\mathbb{R}^2$ consists of a set containing two unit vectors $|b_1\rangle$ and $|b_2\rangle$ that are orthogonal. So, given a pair of kets, to check whether they form an orthonormal basis, we must check first to see if they are unit vectors, and then check whether they are orthogonal. We can check both of these conditions using bra-kets. We need $\langle b_1 | b_1 \rangle = 1$, $\langle b_2 | b_2 \rangle = 1$, and $\langle b_1 | b_2 \rangle = 0$.

The standard example, which is called the *standard basis*, is to take $|b_1\rangle = \begin{bmatrix} 1 \\ 0 \end{bmatrix}$ and $|b_2\rangle = \begin{bmatrix} 0 \\ 1 \end{bmatrix}$. It is straightforward to check that the two bra-ket properties are satisfied. While $\left\{ \begin{bmatrix} 1 \\ 0 \end{bmatrix}, \begin{bmatrix} 0 \\ 1 \end{bmatrix} \right\}$ is a particularly easy basis to find, there are infinitely many other possible choices. Two of these are

$$\left\{ \begin{bmatrix} \dfrac{1}{\sqrt{2}} \\ \dfrac{-1}{\sqrt{2}} \end{bmatrix}, \begin{bmatrix} \dfrac{1}{\sqrt{2}} \\ \dfrac{1}{\sqrt{2}} \end{bmatrix} \right\} \text{ and } \left\{ \begin{bmatrix} \dfrac{1}{2} \\ \dfrac{-\sqrt{3}}{2} \end{bmatrix}, \begin{bmatrix} \dfrac{\sqrt{3}}{2} \\ \dfrac{1}{2} \end{bmatrix} \right\}.$$

In the last chapter we considered measuring the spin of a particle. We looked at spin measured in the vertical direction and in the horizontal direction. The mathematical model for measuring spin in the vertical direction will be given using the standard basis. Rotating the measuring apparatus will be described mathematically by choosing a new orthonormal basis. The three two-dimensional bases* that we have listed will all have important interpretations concerning spin, so instead of naming the vectors in the bases with letters we will use arrows, with the direction of the arrow related to the direction of spin. Here are the names we are going to use:

$$|\uparrow\rangle = \begin{bmatrix} 1 \\ 0 \end{bmatrix}, |\downarrow\rangle = \begin{bmatrix} 0 \\ 1 \end{bmatrix}, |\rightarrow\rangle = \begin{bmatrix} \dfrac{1}{\sqrt{2}} \\ \dfrac{-1}{\sqrt{2}} \end{bmatrix}, |\leftarrow\rangle = \begin{bmatrix} \dfrac{1}{\sqrt{2}} \\ \dfrac{1}{\sqrt{2}} \end{bmatrix}, |\nearrow\rangle = \begin{bmatrix} \dfrac{1}{2} \\ \dfrac{-\sqrt{3}}{2} \end{bmatrix},$$

$$\text{and } |\swarrow\rangle = \begin{bmatrix} \dfrac{\sqrt{3}}{2} \\ \dfrac{1}{2} \end{bmatrix}.$$

* Note that the word *bases* is the plural of *basis* and of *base*. The word is pronounced differently depending on what the singular term is. We will always be using it for the plural of basis. In this case, it is pronounced "BAY-sees."

Our three bases can be written more succinctly as $\{|\uparrow\rangle, |\downarrow\rangle\}$, $\{|\rightarrow\rangle, |\leftarrow\rangle\}$ and $\{|\nearrow\rangle, |\swarrow\rangle\}$. Since these are orthonormal, we have the following bra-ket values.

$$\langle\uparrow|\uparrow\rangle = 1 \quad \langle\downarrow|\downarrow\rangle = 1 \quad \langle\uparrow|\downarrow\rangle = 0 \quad \langle\downarrow|\uparrow\rangle = 0$$
$$\langle\rightarrow|\rightarrow\rangle = 1 \quad \langle\leftarrow|\leftarrow\rangle = 1 \quad \langle\rightarrow|\leftarrow\rangle = 0 \quad \langle\leftarrow|\rightarrow\rangle = 0$$
$$\langle\nearrow|\nearrow\rangle = 1 \quad \langle\swarrow|\swarrow\rangle = 1 \quad \langle\nearrow|\swarrow\rangle = 0 \quad \langle\swarrow|\nearrow\rangle = 0$$

Vectors as Linear Combinations of Basis Vectors

Given a ket and an orthonormal basis, we can express the ket as a weighted sum of the basis vectors. Although at this stage it is not clear that this is useful, we will see later that this is one of the basic ideas on which our mathematical model is based. We start by looking at two-dimensional examples.

Any vector $|v\rangle$ in $\mathbb{R}^2$ can be written as a multiple of $|\uparrow\rangle$ plus a multiple of $|\downarrow\rangle$. This is equivalent to the rather obvious fact that for any numbers c and d the equation

$$\begin{bmatrix} c \\ d \end{bmatrix} = x_1 \begin{bmatrix} 1 \\ 0 \end{bmatrix} + x_2 \begin{bmatrix} 0 \\ 1 \end{bmatrix}$$

has a solution. Clearly, this has a solution of $x_1 = c$ and $x_2 = d$, and this is the only solution.

Can any vector $|v\rangle$ in $\mathbb{R}^2$ be written as a multiple of $|\rightarrow\rangle$ plus a multiple of $|\leftarrow\rangle$? Equivalently, does the following equation have a solution for any numbers c and d?

$$\begin{bmatrix} c \\ d \end{bmatrix} = x_1 |\rightarrow\rangle + x_2 |\leftarrow\rangle.$$

How do we solve this? We could replace the kets with their two-dimensional column vectors and then solve the resulting system of two linear equations in two unknowns. But there is a far easier way of doing this using bras and kets.

First, take the equation and multiply both sides on the left by the bra $\langle\rightarrow|$ This gives us the following equation.

$$\langle\rightarrow| \begin{bmatrix} c \\ d \end{bmatrix} = \langle\rightarrow|(x_1|\rightarrow\rangle + x_2|\leftarrow\rangle)$$

Next, distribute the terms on the right side of the equation.

$$\langle\rightarrow|\begin{bmatrix}c\\d\end{bmatrix}=x_1\langle\rightarrow|\rightarrow\rangle+x_2\langle\rightarrow|\leftarrow\rangle$$

We know both of the bra-kets on the right side. The first is 1. The second is 0. This immediately tells us that x_1 is equal to $\langle\rightarrow|\begin{bmatrix}c\\d\end{bmatrix}$. So, we just need to evaluate this product.

$$\langle\rightarrow|\begin{bmatrix}c\\d\end{bmatrix}=\begin{bmatrix}1/\sqrt{2} & -1/\sqrt{2}\end{bmatrix}\begin{bmatrix}c\\d\end{bmatrix}=\left(1/\sqrt{2}\right)c-\left(1/\sqrt{2}\right)d=(c-d)/\sqrt{2}.$$

Consequently, $x_1=(c-d)/\sqrt{2}$.

We can use exactly the same method to find x_2. We start with the same initial equation $\begin{bmatrix}c\\d\end{bmatrix}=x_1|\rightarrow\rangle+x_2|\leftarrow\rangle$ and multiply both sides on the left by the bra $\langle\leftarrow|$.

$$\langle\leftarrow|\begin{bmatrix}c\\d\end{bmatrix}=x_1\langle\leftarrow|\rightarrow\rangle+x_2\langle\leftarrow|\leftarrow\rangle=x_10+x_21.$$

So, $x_2=\begin{bmatrix}1/\sqrt{2} & 1/\sqrt{2}\end{bmatrix}\begin{bmatrix}c\\d\end{bmatrix}=\left(1/\sqrt{2}\right)c+\left(1/\sqrt{2}\right)d=(c+d)/\sqrt{2}.$

This means that we can write

$$\begin{bmatrix}c\\d\end{bmatrix}=\frac{(c-d)}{\sqrt{2}}|\rightarrow\rangle+\frac{(c+d)}{\sqrt{2}}|\leftarrow\rangle.$$

The sum on the right consists of multiplying the basis vectors by certain scalars and then adding the resulting vectors. I described it earlier as a weighted sum of the basis vectors, but you have to be careful with this interpretation. There is no reason for the scalars to be positive. They can be negative. In our example, if c were to equal –3 and d were to equal 1, both of the weights, $(c-d)/\sqrt{2}$ and $(c+d)/\sqrt{2}$, would be negative. For this reason the term *linear combination* of the basis vectors is used instead of weighted sum.

Now let's move to n dimensions. Suppose that we are given an n-dimensional ket $|v\rangle$ and an orthonormal basis $\{|b_1\rangle,|b_2\rangle,\cdots,|b_n\rangle\}$. Can we write $|v\rangle$ as a linear combination of the basis vectors? If so, is there a unique way of doing this? Equivalently, does the equation

$$|v\rangle = x_1 |b_1\rangle + x_2 |b_2\rangle + \cdots + x_i |b_i\rangle + \cdots + x_n |b_n\rangle$$

have a unique solution? Again, the answer is yes. To see this we will show how to find the value for x_i. The calculation follows exactly the same method we used in two dimensions. Start by multiplying both sides of the equation by $\langle b_i |$. We know that $\langle b_i | b_k \rangle$ equals 0 if $i \neq k$ and equals 1 if $i = k$. So, after multiplying by the bra, the right side simplifies to just x_i, and we obtain that $\langle b_i | v \rangle = x_i$. This tells us that $x_1 = \langle b_1 | v \rangle$, $x_2 = \langle b_2 | v \rangle$, etc. Consequently, we can write $| v \rangle$ as a linear combination of the basis vectors:

$$|v\rangle = \langle b_1 |v\rangle |b_1\rangle + \langle b_2 |v\rangle |b_2\rangle + \cdots + \langle b_i |v\rangle |b_i\rangle + \cdots + \langle b_n |v\rangle |b_n\rangle$$

At this stage, this all seems somewhat abstract, but it will all become clear in the next chapter. Different orthonormal bases correspond to choosing different orientations to measure spin. The numbers given by the brakets like $\langle b_i | v \rangle$ are called *probability amplitudes*. The square of $\langle b_i | v \rangle$ will give us the probability of $|v\rangle$ jumping to $|b_i\rangle$ when we measure it. This will all be explained, but understanding the equation written above is crucial to what follows.

Ordered Bases

An *ordered* basis is a basis in which the vectors have been given an order, that is, there is a first vector, a second vector, and so on. If $\{|b_1\rangle, |b_2\rangle, \cdots, |b_n\rangle\}$ is a basis, we will denote the ordered basis by $(|b_1\rangle, |b_2\rangle, \cdots, |b_n\rangle)$—we change the brackets from curly to round. For an example, we will look at $\mathbb{R}^2$. Recall that the standard basis is $\{|\uparrow\rangle, |\downarrow\rangle\}$. Two sets are equal if they have the same elements—the order of the elements does not matter, so $\{|\uparrow\rangle, |\downarrow\rangle\} = \{|\downarrow\rangle, |\uparrow\rangle\}$. The two sets are identical.

However, for an ordered basis the order the basis vectors are given matters. $(|\uparrow\rangle, |\downarrow\rangle) \neq (|\downarrow\rangle, |\uparrow\rangle)$. The first vector in the ordered basis on the left is not equal to the first vector in the ordered basis on the right, so the two ordered bases are distinct.

The difference between unordered bases and ordered bases might seem rather pedantic, but it is not. We will see several examples where we have the same set of basis vectors in which the order is different. The permutation of the basis vectors will give us important information.

As an example, earlier we noted that the standard basis $\{|\uparrow\rangle, |\downarrow\rangle\}$ corresponds to measuring the spin of an electron in the vertical direction. The ordered basis $(|\uparrow\rangle, |\downarrow\rangle)$ will correspond to measuring the spin when the south magnet is on top of our measuring apparatus. If we flip the apparatus through 180°, we will also flip the basis elements and use the ordered basis $(|\downarrow\rangle, |\uparrow\rangle)$.

Length of Vectors

Supposing that we have been given a ket $|v\rangle$ and an orthonormal basis $\{|b_1\rangle, |b_2\rangle, \cdots, |b_n\rangle\}$, we know how to write $|v\rangle$ as a linear combination of the basis vectors. We end up with $|v\rangle = \langle b_1|v\rangle|b_1\rangle + \langle b_2|v\rangle|b_2\rangle + \cdots + \langle b_i|v\rangle|b_i\rangle + \cdots + \langle b_n|v\rangle|b_n\rangle$. To simplify things, we will write this as $|v\rangle = c_1|b_1\rangle + c_2|b_2\rangle + \cdots + c_i|b_i\rangle + \cdots + c_n|b_n\rangle$. There is a useful formula for the length of $|v\rangle$. It's $\||v\rangle\|^2 = c_1^2 + c_2^2 + \cdots + c_i^2 \cdots + \cdots c_n^2$.

Let's quickly see why this is true. We know that $\||v\rangle\|^2 = \langle v|v\rangle$.

Using $\langle v| = c_1\langle b_1| + c_2\langle b_2| + \cdots + c_n\langle b_n|$ we obtain

$$\langle v|v\rangle = (c_1\langle b_1| + c_2\langle b_2| + \cdots + c_n\langle b_n|)(c_1|b_1\rangle + c_2|b_2\rangle + \cdots + c_n|b_n\rangle).$$

The next step is to expand the product of the terms in the parentheses. This looks as though it is going to be messy, but it is not. We again use the facts that $\langle b_i|b_k\rangle$ equals 0 if $i \neq k$ and equals 1 if $i = k$. All the bra-ket products with different subscripts are 0. The only bra-kets that are nonzero are the ones where the same subscript is repeated, and these are all 1. Consequently, we end up with $\langle v|v\rangle = c_1^2 + c_2^2 + \cdots + c_i^2 + \cdots c_n^2$.

Matrices

Matrices are rectangular arrays of numbers. A matrix M with m rows and n columns is called an $m \times n$ matrix. Here are a couple of examples:

$$A = \begin{bmatrix} 1 & -4 & 2 \\ 2 & 3 & 0 \end{bmatrix} \quad B = \begin{bmatrix} 1 & 2 \\ 7 & 5 \\ 6 & 1 \end{bmatrix}$$

A has two rows and three columns so it is a 2×3 matrix. B is a 3×2 matrix. We can think of bras and kets as being special types of matrices: bras have just one row, and kets have just one column.

The *transpose* of a $m \times n$ matrix M, denoted M^T, is the $n \times m$ matrix formed by interchanging the rows and the columns of M. The ith row of M becomes the ith column of M^T, and the jth column of M becomes the jth row of M^T. For our matrices A and B we have:

$$A^T = \begin{bmatrix} 1 & 2 \\ -4 & 3 \\ 2 & 0 \end{bmatrix} B^T = \begin{bmatrix} 1 & 7 & 6 \\ 2 & 5 & 1 \end{bmatrix}$$

Column vectors can be considered as matrices with just one column, and row vectors can be considered as matrices with just one row. With this interpretation, the relation between bras and kets with the same name is given by $\langle a | = | a \rangle^T$ and $| a \rangle = \langle a |^T$.

Given a general matrix that has multiple rows and columns, we think of the rows as denoting bras and the columns as denoting kets. In our example, we can think of A as consisting of two bras stacked on one another or as three kets side by side. Similarly, B can be considered as three bras stacked on one another or as two kets side by side.

The product of the matrices A and B uses this idea. The product is denoted by AB. It's calculated by thinking of A as consisting of bras and B of kets. (Remember that bras always come before kets.)

$$A = \begin{bmatrix} \langle a_1 | \\ \langle a_2 | \end{bmatrix}, \text{ where } \langle a_1 | = [1 \quad -4 \quad 2] \text{ and } \langle a_2 | = [2 \quad 3 \quad 0].$$

$$B = [| b_1 \rangle \quad | b_2 \rangle], \text{ where } | b_1 \rangle = \begin{bmatrix} 1 \\ 7 \\ 6 \end{bmatrix} \text{ and } | b_2 \rangle = \begin{bmatrix} 2 \\ 5 \\ 1 \end{bmatrix}.$$

The product AB is calculated as follows:

$$AB = \begin{bmatrix} \langle a_1 | \\ \langle a_2 | \end{bmatrix} [| b_1 \rangle \quad | b_2 \rangle] = \begin{bmatrix} \langle a_1 | b_1 \rangle & \langle a_1 | b_2 \rangle \\ \langle a_2 | b_1 \rangle & \langle a_2 | b_2 \rangle \end{bmatrix}$$

$$= \begin{bmatrix} 1 \times 1 - 4 \times 7 + 2 \times 6 & 1 \times 2 - 4 \times 5 + 2 \times 1 \\ 2 \times 1 + 3 \times 7 + 0 \times 6 & 2 \times 2 + 3 \times 5 + 0 \times 1 \end{bmatrix}$$

$$= \begin{bmatrix} -15 & -16 \\ 23 & 19 \end{bmatrix}$$

Notice that the dimension of the bras in A is equal to the dimension of the kets in B. We need to have this in order for the bra-ket products to be defined. Also notice that $AB \neq BA$. In our example, BA is a 3×3 matrix, so it is not even the same size as AB.

In general, given an $m \times r$ matrix A and an $r \times n$ matrix B, write A in terms of r-dimensional bras and B in terms of r-dimensional kets.

$$A = \begin{bmatrix} \langle a_1 | \\ \langle a_2 | \\ \vdots \\ \langle a_m | \end{bmatrix} \quad B = [| b_1 \rangle \quad | b_2 \rangle \quad \cdots \quad | b_n \rangle],$$

The product AB is the $m \times n$ matrix that has $\langle a_i | b_j \rangle$ as the entry in the ith row and jth column, that is,

$$AB = \begin{bmatrix} \langle a_1|b_1 \rangle & \langle a_1|b_2 \rangle & \cdots & \langle a_1|b_j \rangle & \cdots & \langle a_1|b_n \rangle \\ \langle a_2|b_1 \rangle & \langle a_2|b_2 \rangle & \cdots & \langle a_2|b_j \rangle & \cdots & \langle a_2|b_n \rangle \\ \vdots & \vdots & \vdots & \vdots & \vdots & \vdots \\ \langle a_i|b_1 \rangle & \langle a_i|b_2 \rangle & \cdots & \langle a_i|b_j \rangle & \cdots & \langle a_i|b_n \rangle \\ \vdots & \vdots & \vdots & \vdots & \vdots & \vdots \\ \langle a_m|b_1 \rangle & \langle a_m|b_2 \rangle & \cdots & \langle a_m|b_j \rangle & \cdots & \langle a_m|b_n \rangle \end{bmatrix}$$

Reversing the order of multiplication gives BA, but we cannot even begin the calculation if m is not equal to n because the bras and kets would have different dimensions. Even if m is equal to n, and we can multiply them, we would end up with a matrix that has size $r \times r$. This is not equal to AB, which has size $n \times n$, if n is not equal to r. Even in the case when n, m and r are all equal to one another, it is usually not the case that AB will equal BA. We say that matrix multiplication is *not commutative* to indicate this fact.

Matrices with the same number of rows as columns are called *square* matrices. The *main diagonal* of a square matrix consists of the elements on the diagonal going from the top left of the matrix to the bottom right. A square matrix that has all leading diagonal entries equal to 1 and all other entries equal to 0 is called an *identity* matrix. The $n \times n$ identity matrix is denoted by I_n.

$$I_2 = \begin{bmatrix} 1 & 0 \\ 0 & 1 \end{bmatrix}, \quad I_3 = \begin{bmatrix} 1 & 0 & 0 \\ 0 & 1 & 0 \\ 0 & 0 & 1 \end{bmatrix}, \quad \ldots$$

The identity matrix gets its name from the fact that multiplying matrices by the identity is analogous to multiplying numbers by 1. Suppose that A is an $m \times n$ matrix. Then $I_m A = AI_n = A$.

Matrices give us a convenient way of doing computations that involve bras and kets. The next section shows how we will be using them.

Matrix Computations

Suppose that we are given a set of n-dimensional kets $\{|b_1\rangle, |b_2\rangle, \cdots, |b_n\rangle\}$ and we want to check to see if it is an orthonormal basis. First, we have to check that they are all unit vectors. Then we have to check that the vectors are mutually orthogonal to one another. We have seen how to check both of these conditions using bras and kets, but the calculation can be expressed simply using matrices.

We begin by forming the $n \times n$ matrix $A = [|b_1\rangle \quad |b_2\rangle \quad \cdots \quad |b_n\rangle]$, then take its transpose.

$$A^T = \begin{bmatrix} \langle b_1 | \\ \langle b_2 | \\ \vdots \\ \langle b_n | \end{bmatrix}$$

Then we take the product $A^T A$.

$$A^T A = \begin{bmatrix} \langle b_1 | \\ \langle b_2 | \\ \vdots \\ \langle b_n | \end{bmatrix} [|b_1\rangle \quad |b_2\rangle \quad \cdots \quad |b_n\rangle] = \begin{bmatrix} \langle b_1|b_1\rangle & \langle b_1|b_2\rangle & \cdots & \langle b_1|b_n\rangle \\ \langle b_2|b_1\rangle & \langle b_2|b_2\rangle & \cdots & \langle b_2|b_n\rangle \\ \vdots & \vdots & \vdots & \vdots \\ \langle b_n|b_1\rangle & \langle b_n|b_2\rangle & \cdots & \langle b_n|b_n\rangle \end{bmatrix}$$

Notice that the entries down the main diagonal are exactly what we need to calculate in order to find if the kets are unit. And the entries off the diagonal are what we have to calculate to see if the kets are mutually orthogonal. This means that the set of vectors is an orthonormal basis if and only if $A^T A = I_n$. This equation gives a succinct way of writing down everything that we need to check.

Though it is a concise expression, we still need to do all the calculations to find the entries. We need to calculate all the entries along the main diagonal in order to check that the vectors are unit. However, we don't need to calculate the entries below the main diagonal. If $i \neq j$ then one of $\langle b_i | b_k \rangle$

and $\langle b_k \mid b_i \rangle$ will be above and the other below the main diagonal. These two bra-ket products are equal, and once we have calculated one we don't need to calculate the other. So, after we have checked that all the main diagonal entries are 1, we just need to check that all the entries above (or below) the diagonal are 0.

Now that we have checked that $\{|b_1\rangle, |b_2\rangle, \cdots, |b_n\rangle\}$ is an orthonormal basis, suppose that we are given a ket $|v\rangle$ and want to express it as a linear combination of the basis vectors. We know how to do this.

$$|v\rangle = \langle b_1 |v\rangle |b_1\rangle + \langle b_2 |v\rangle |b_2\rangle + \cdots + \langle b_i |v\rangle |b_i\rangle + \cdots + \langle b_n |v\rangle |b_n\rangle$$

Everything can be calculated using the matrix A^T.

$$A^T |v\rangle = \begin{bmatrix} \langle b_1 | \\ \langle b_2 | \\ \vdots \\ \langle b_n | \end{bmatrix} |v\rangle = \begin{bmatrix} \langle b_1 | v\rangle \\ \langle b_2 | v\rangle \\ \vdots \\ \langle b_n | v\rangle \end{bmatrix}$$

This has been a long chapter in which much mathematical machinery has been introduced. But the mathematics has been building, and we now have a number of ways for performing calculations. Three key ideas that we will need later are summarized in the final section. (They are at the end of the chapter for easy reference.) Before we conclude we look at some naming conventions.

Orthogonal and Unitary Matrices

A square matrix M that has real entries and has the property that $M^T M$ is equal to the identity matrix is called an *orthogonal* matrix.

As we saw in the last section, we can check to see whether we have an orthonormal basis by forming the matrix of the kets and then checking whether the resulting matrix is orthogonal. Orthogonal matrices will also be important when we look at quantum logic gates. These gates also correspond to orthogonal matrices.

Two important orthogonal matrices are

$$\begin{bmatrix} \dfrac{1}{\sqrt{2}} & \dfrac{1}{\sqrt{2}} \\ \dfrac{1}{\sqrt{2}} & \dfrac{-1}{\sqrt{2}} \end{bmatrix} \text{ and } \begin{bmatrix} 1 & 0 & 0 & 0 \\ 0 & 1 & 0 & 0 \\ 0 & 0 & 0 & 1 \\ 0 & 0 & 1 & 0 \end{bmatrix}.$$

The 2×2 matrix corresponds to the ordered basis $(|\leftarrow\rangle, |\rightarrow\rangle)$, which we will meet in the next chapter where we will see how it is connected to measuring spin in the horizontal direction. We will also meet exactly the same matrix later. It is the matrix corresponding to a special gate, called the *Hadamard* gate.

The 4×4 matrix corresponds to taking the standard basis for $\mathbb{R}^4$ and ordering with the last two vectors interchanged. This matrix is associated with the *CNOT* gate. We will explain later exactly what gates are, but practically all of our quantum circuits will be composed of just these two types of gates. So, these orthogonal matrices are important!

(If we were working with complex numbers, the matrix entries could be complex numbers. Matrices with complex entries that correspond to orthogonal matrices are called *unitary*.** Real numbers are a subset of the complex numbers, so all orthogonal matrices are unitary. If you look at practically every other book on quantum computing, they will call the matrices describing the *CNOT* gate and the Hadamard gate unitary, but we are calling them orthogonal. Both are correct.)

Linear Algebra Toolbox

Here is a list of three tasks that we will need to perform repeatedly. These are all easy to do. The methods for tackling each task are given.

(1) Given a set of n-dimensional kets $\{|b_1\rangle, |b_2\rangle, \cdots, |b_n\rangle\}$, check to see if it is an orthonormal basis.

To do this, first construct $A = [|b_1\rangle \ |b_2\rangle \ \cdots \ |b_n\rangle]$. Then compute $A^T A$. If this is the identity matrix, we have an orthonormal basis. If it isn't, we don't.

(2) Given an orthonormal basis $\{|b_1\rangle, |b_2\rangle, \cdots, |b_n\rangle\}$ and a ket $|v\rangle$, express the ket as a linear combination of the basis vectors, that is, solve

$$|v\rangle = x_1 |b_1\rangle + x_2 |b_2\rangle + \cdots x_i |b_i\rangle \cdots + x_n |b_n\rangle.$$

To do this, construct $A = [|b_1\rangle \ |b_2\rangle \ \cdots \ |b_n\rangle]$. Then

** A matrix M is unitary if $M^\dagger M$ is the identity matrix, where $M^\dagger$ means first transpose M, then take the conjugate of all the entries.

$$\begin{bmatrix} x_1 \\ x_2 \\ \vdots \\ x_n \end{bmatrix} = A^T |v\rangle = \begin{bmatrix} \langle b_1 | v \rangle \\ \langle b_2 | v \rangle \\ \vdots \\ \langle b_n | v \rangle \end{bmatrix}$$

(3) Given an orthonormal basis $\{|b_1\rangle, |b_2\rangle, \cdots, |b_n\rangle\}$ and

$|v\rangle = c_1 |b_1\rangle + c_2 |b_2\rangle + \cdots c_i |b_i\rangle \cdots + c_n |b_n\rangle,$ find the length of $|v\rangle$.

To do this, use $\||v\rangle\|^2 = c_1^2 + c_2^2 + \cdots + c_i^2 \cdots + \cdots c_n^2.$

Now that we have the tools, we return to the study of spin.

3   Spin and Qubits

The first chapter described measurements involving the spin of an electron. We saw that if you measure the spin in the vertical direction, you don't obtain a continuum of values, but just two of them: Either the electron has its north pole vertically upward, or it is vertically downward. If we measure the spin first in the vertical direction and then once more in the same direction, we obtain exactly the same result for both measurements. If the first measurement shows that the electron has its north pole upward, then so will the second measurement. We also saw that if we measure first in the vertical direction and then in the horizontal direction, the electrons will have spin N and S in direction 90° each with probability of one half. It doesn't matter what the first measurement is; the second measurement will give a random choice of either N or S. The second chapter introduced the mathematics of linear algebra. The goal of this chapter is to combine these previous two chapters, giving a mathematical model that describes the measurement of spin. We will then show how this relates to qubits. But before we start this description, we introduce the mathematics of probability.

Probability

Imagine that we have a coin and we repeatedly toss it, counting both the number of tosses and the number of times it comes up heads. If the coin is fair—equally likely to land heads up as tails up—the ratio of the number of heads to the number of tosses, after tossing it a large number of times, will be close to one half. We say that the probability of the outcome "heads" is 0.5.

In general, we perform an experiment—often we will call it making a measurement—that has a finite number of possible outcomes. We will denote these by $E_1, E_2, \ldots, E_n$. The underlying assumption is that the result of the experiment, or measurement, will be one and only one of these n outcomes. Associated with outcome E_i is a *probability* p_i. Probabilities must be numbers between 0 and 1 that sum to 1. In the case of tossing a coin, the two outcomes are getting a head and getting a tail. If the coin is fair, the probability of each event is $1/2$.

We return to the experiments involving the spin of a particle from the first chapter using a slightly more formal notation to describe them. Suppose that we are going to measure the spin in direction $0°$. There are two possible outcomes that we will denote as N and S. Both of these outcomes will have an associated probability. We will denote by p_N the probability of obtaining N, and p_S the probability of obtaining S. If we already know that our electron has spin N in direction $0°$, then we know that when we measure again in this direction we will get the same result, so, in this case, $p_N = 1$ and $p_S = 0$. On the other hand, if we know our electron has spin N in direction $90°$ and we now measure in direction $0°$, then we are equally likely to obtain N and S as the outcome, so, in this case $p_N = p_S = 0.5$.

Mathematics of Quantum Spin

We will now present the mathematical model that describes quantum spin. It uses both probabilities and vectors.

The basic model is given by a vector space. When we make a measurement there will be a number of possible outcomes. The number of outcomes determines the dimension of this underlying vector space. For spin, there are just two possible outcomes from any measurement, so the underlying vector space is two-dimensional. We will take the space to be $\mathbb{R}^2$—this is the standard two-dimensional plane with which we are all familiar. This is fine for our purposes because we are only rotating our measuring apparatus in the plane. If we also wanted to consider all possible three-dimensional rotations of the apparatus, the underlying space would still be two-dimensional—two is still the number of possible outcomes for each measurement—but instead of using vectors with real coefficients, we would have to use vectors that involve complex numbers. The underlying vector space would then be the two-dimensional complex space denoted $\mathbb{C}^2$. For the reasons listed in the previous chapter, $\mathbb{R}^2$ is fine for our needs.

We will not consider all of the vectors in $\mathbb{R}^2$, just the unit vectors. For kets, this means we are restricting to kets of the form $|v\rangle = \begin{bmatrix} c_1 \\ c_2 \end{bmatrix}$, where $c_1^2 + c_2^2 = 1$.

Choosing a direction to measure spin corresponds to choosing an ordered, orthonormal basis $(|b_1\rangle, |b_2\rangle)$. The two vectors in the basis correspond to the two possible outcomes for the measurements. We will always associate N with the first basis vector and S with the second. Before we measure the spin, the particle will be in a *spin state* given by a linear combination of $|b_1\rangle$ and $|b_2\rangle$, that is, it has the form $c_1|b_1\rangle + c_2|b_2\rangle$. We will sometimes refer to this as a *state vector* and sometimes just call it a *state*. After we measure, its state vector will jump to either $|b_1\rangle$ or $|b_2\rangle$. This is one of the major ideas in quantum mechanics: Measurement causes the state vector to change. The new state is one of the basis vectors associated with the measurement. The probability of getting a particular basis vector is given by the initial state. The probability of its being $|b_1\rangle$ is c_1^2; the probability of $|b_2\rangle$ is c_2^2. The numbers c_1 and c_2 are called the *probability amplitudes*. It's important to remember that the probability amplitudes are not probabilities. They can be positive or negative. It's the squares of these numbers that are probabilities. To make everything more concrete, we will return to the experiments where we measured spin in the vertical and horizontal directions.

As we mentioned in the previous chapter, the ordered orthonormal basis corresponding to measuring spin in the vertical direction is given by $(|\uparrow\rangle, |\downarrow\rangle)$, where $|\uparrow\rangle = \begin{bmatrix} 1 \\ 0 \end{bmatrix}$ and $|\downarrow\rangle = \begin{bmatrix} 0 \\ 1 \end{bmatrix}$. The first vector listed in the basis corresponds to the electron having spin N in direction $0°$ and the second vector to S in direction $0°$.

The spin in the horizontal direction is given by the ordered orthonormal basis $(|\rightarrow\rangle, |\leftarrow\rangle)$, where $|\rightarrow\rangle = \begin{bmatrix} \frac{1}{\sqrt{2}} \\ \frac{-1}{\sqrt{2}} \end{bmatrix}$ and $|\leftarrow\rangle = \begin{bmatrix} \frac{1}{\sqrt{2}} \\ \frac{1}{\sqrt{2}} \end{bmatrix}$. The first vector listed in the basis corresponds to the electron having spin N in direction $90°$ and the second vector to S in direction $90°$.

We first measure spin in the vertical direction. Initially, we might not know the spin state of the incoming electron, but it must be a unit vector and so can be written as $c_1|\uparrow\rangle + c_2|\downarrow\rangle$, where $c_1^2 + c_2^2 = 1$. We now perform the measurement. Either the electron is diverted upward in which case the state jumps to $|\uparrow\rangle$ or it is diverted downward in which case its state jumps to $|\downarrow\rangle$. The probability of it being diverted upward is c_1^2 and the probability of it being diverted downward is c_2^2.

We now repeat exactly the same experiment, measuring the spin once more in the vertical direction. Suppose that the electron was deflected upward by the first set of magnets. We know it is in spin state $|\uparrow\rangle = 1|\uparrow\rangle + 0|\downarrow\rangle$. When we measure again, the state jumps to $|\uparrow\rangle$ with probability $1^2 = 1$, or to $|\downarrow\rangle$ with probability $0^2 = 0$. This just means that it remains in state $|\uparrow\rangle$, and so is deflected upward once more.

Similarly, if the electron was deflected downward it will be in state $|\downarrow\rangle = 0|\uparrow\rangle + 1|\downarrow\rangle$. No matter how many times we measure it in the vertical direction it will remain in this state, telling us that however many times we repeat the experiment the electron will keep being deflected downward. As we noted in the first chapter, if we repeat exactly the same experiment, we get exactly the same outcome.

Instead of repeatedly measuring spin in the vertical direction, we will first measure spin in the vertical direction and then measure spin in the horizontal direction. Suppose that we have just performed the first measurement. We have measured spin in the vertical direction, and let us suppose that the electron has spin N in direction $0°$. Its state vector is now $|\uparrow\rangle$. Since we are next going to measure spin in the horizontal direction, we have to write this vector in terms of the orthonormal basis that corresponds to this direction, which means we must find the values of x_1 and x_2 that solve $|\uparrow\rangle = x_1|\rightarrow\rangle + x_2|\leftarrow\rangle$. We know how to do this: It's the second tool in the toolbox listed at the end of the last chapter.

First construct the matrix A by stacking the kets that form the orthonormal basis side by side.

$$A = [|\rightarrow\rangle|\leftarrow\rangle] = \begin{bmatrix} \dfrac{1}{\sqrt{2}} & \dfrac{1}{\sqrt{2}} \\ \dfrac{-1}{\sqrt{2}} & \dfrac{1}{\sqrt{2}} \end{bmatrix}$$

Then calculate $A^T|\uparrow\rangle$ to get the probability amplitudes with respect to the new basis.

$$A^T |\uparrow\rangle = \begin{bmatrix} \dfrac{1}{\sqrt{2}} & \dfrac{-1}{\sqrt{2}} \\ \dfrac{1}{\sqrt{2}} & \dfrac{1}{\sqrt{2}} \end{bmatrix} \begin{bmatrix} 1 \\ 0 \end{bmatrix} = \begin{bmatrix} \dfrac{1}{\sqrt{2}} \\ \dfrac{1}{\sqrt{2}} \end{bmatrix}$$

This tells us that $|\uparrow\rangle = \dfrac{1}{\sqrt{2}} |\rightarrow\rangle + \dfrac{1}{\sqrt{2}} |\leftarrow\rangle$.

When we measure in the horizontal direction the state will jump to $|\rightarrow\rangle$ with probability $\left(\dfrac{1}{\sqrt{2}}\right)^2 = \dfrac{1}{2}$, or it will jump to $|\leftarrow\rangle$ with probability $\left(\dfrac{1}{\sqrt{2}}\right)^2 = \dfrac{1}{2}$. This tells us that the probability that the electron has spin N in the 90° direction is equal to the probability that it has spin S in the 90°direction; both probabilities are exactly one-half.

Notice that we didn't really need to calculate the matrix A to do this calculation. The matrix that we need to use is A^T. We can calculate this by stacking the bras that correspond to the orthonormal basis on top of one another. We must, of course, keep things in the same order. The left to right ordering of kets corresponds to the top to bottom ordering of bras, so the first element of the basis is the topmost bra.

In the first chapter we measured the spin three times. The first and third measurements were in the vertical direction, the second was in the horizontal direction. We will describe the mathematics that corresponds to the third measurement. After the second measurement, the state vector of our electron will have one of two values. It will be either $|\rightarrow\rangle$ or $|\leftarrow\rangle$. We are now going to measure the spin in the vertical direction, so we need to express these as linear combinations of the vertical orthonormal basis. This gives $|\rightarrow\rangle = \dfrac{1}{\sqrt{2}} |\uparrow\rangle - \dfrac{1}{\sqrt{2}} |\downarrow\rangle$ and $|\leftarrow\rangle = \dfrac{1}{\sqrt{2}} |\uparrow\rangle + \dfrac{1}{\sqrt{2}} |\downarrow\rangle$. In either case, when we measure spin in the vertical direction the state vector will jump to either $|\uparrow\rangle$ or to $|\downarrow\rangle$, each occurring with probability one-half.

Equivalent State Vectors

Suppose that we are given a number of electrons and are told that their spins are given by either $|\uparrow\rangle$ or by $-|\uparrow\rangle$. Can we distinguish between the two cases? Is there any measurement that we can perform that would tell them apart? The answer is that there is not.

To see this, let's suppose that we choose a direction in which to measure spin. This is equivalent to choosing an ordered, orthonormal basis. We will denote this basis by $(|b_1\rangle, |b_2\rangle)$.

Suppose our electron has state $|\uparrow\rangle$. We have to find the values of a and b that solve the equation $|\uparrow\rangle = a|b_1\rangle + b|b_2\rangle$. When we perform the measurement, the probability of the spin being N is a^2, and the probability of the spin being S is b^2.

Suppose our electron has state $-|\uparrow\rangle$. For exactly the same values of a and b, we have $-|\uparrow\rangle = -a|b_1\rangle - b|b_2\rangle$. When we perform the measurement the probability of the spin being N is $(-a)^2 = a^2$ and the probability of the spin being S is $(-b)^2 = b^2$.

We get exactly the same probabilities for both cases, so there is no measurement that can distinguish electrons with state vectors of form $|\uparrow\rangle$ from those of $-|\uparrow\rangle$.

Similarly, given electrons with state $|v\rangle$ there is no way to distinguish them from electrons with state $-|v\rangle$. Since these states are indistinguishable, they are considered equivalent. Saying that an electron has spin given by $|v\rangle$ means exactly the same as saying that it has spin given by $-|v\rangle$.

To help illustrate this point further, consider these four kets:

$$\frac{1}{\sqrt{2}}|\uparrow\rangle + \frac{1}{\sqrt{2}}|\downarrow\rangle \qquad -\frac{1}{\sqrt{2}}|\uparrow\rangle - \frac{1}{\sqrt{2}}|\downarrow\rangle \qquad \frac{1}{\sqrt{2}}|\uparrow\rangle - \frac{1}{\sqrt{2}}|\downarrow\rangle \qquad -\frac{1}{\sqrt{2}}|\uparrow\rangle + \frac{1}{\sqrt{2}}|\downarrow\rangle$$

By the preceding remarks, we know that $\frac{1}{\sqrt{2}}|\uparrow\rangle + \frac{1}{\sqrt{2}}|\downarrow\rangle$ and $-\frac{1}{\sqrt{2}}|\uparrow\rangle - \frac{1}{\sqrt{2}}|\downarrow\rangle$ are equivalent, and that $\frac{1}{\sqrt{2}}|\uparrow\rangle - \frac{1}{\sqrt{2}}|\downarrow\rangle$ and $-\frac{1}{\sqrt{2}}|\uparrow\rangle + \frac{1}{\sqrt{2}}|\downarrow\rangle$ are equivalent. So, these four kets describe at most two distinguishable states. But what about $\frac{1}{\sqrt{2}}|\uparrow\rangle + \frac{1}{\sqrt{2}}|\downarrow\rangle$ and $\frac{1}{\sqrt{2}}|\uparrow\rangle - \frac{1}{\sqrt{2}}|\downarrow\rangle$? Do these describe the same state, or are they distinguishable?

We do have to be a little careful. If we choose to measure the spin in the vertical direction, these two kets are not distinguishable. In both cases, we get $|\uparrow\rangle$ or $|\downarrow\rangle$ each occurring with probability of a half. But we know that $\frac{1}{\sqrt{2}}|\uparrow\rangle + \frac{1}{\sqrt{2}}|\downarrow\rangle = |\leftarrow\rangle$ and $\frac{1}{\sqrt{2}}|\uparrow\rangle - \frac{1}{\sqrt{2}}|\downarrow\rangle = |\rightarrow\rangle$. Consequently, if we measure in the 90° direction, we will obtain S for the first ket and N for the second. This choice of basis does distinguish them, and so they are not equivalent.

One thing that is probably not clear at the moment is how the basis associated with a direction of measurement is chosen. We have seen that the basis associated with measuring in the vertical (0°) direction is $\left(\begin{bmatrix} 1 \\ 0 \end{bmatrix}, \begin{bmatrix} 0 \\ 1 \end{bmatrix}\right)$ and with the horizontal (90°) direction is $\left(\begin{bmatrix} \frac{1}{\sqrt{2}} \\ \frac{-1}{\sqrt{2}} \end{bmatrix}, \begin{bmatrix} \frac{1}{\sqrt{2}} \\ \frac{1}{\sqrt{2}} \end{bmatrix}\right)$.

But where did these bases come from? Later, when we come to Bell's theorem, we will need the bases associated with 120° and 240°. What are these? We answer these questions in the next section.

The Basis Associated with a Given Spin Direction

We begin with our measurement apparatus. We take the vertical direction as the starting point and start rotating in the clockwise direction. As we have already noted, when it has been rotated through 90°, we are measuring in the horizontal direction. By the time it's rotated through 180°, we are measuring the vertical direction once more. An electron that has spin N in direction 0° will have spin S in direction 180°, and an electron that has spin S in direction 0° will have spin N in direction 180°. Clearly, saying a magnet has its north pole in one direction conveys exactly the same information as saying the magnet has its south pole in the opposite direction, and consequently we need only to rotate our apparatus through an angle between 0° and 180° to cover all possible directions.

We will now consider bases. We take the standard basis $\left(\begin{bmatrix} 1 \\ 0 \end{bmatrix}, \begin{bmatrix} 0 \\ 1 \end{bmatrix}\right)$ as our starting point. This can be pictured as two vectors in the plane, as shown in figure 3.1.

Now we rotate these vectors. The general picture, with rotation of $\alpha°$ is depicted in figure 3.2. The vector $\begin{bmatrix} 1 \\ 0 \end{bmatrix}$ rotates to $\begin{bmatrix} \cos(\alpha) \\ -\sin(\alpha) \end{bmatrix}$, and $\begin{bmatrix} 0 \\ 1 \end{bmatrix}$ rotates to $\begin{bmatrix} \sin(\alpha) \\ \cos(\alpha) \end{bmatrix}$.

Rotating through $\alpha°$ changes our initial ordered, orthonormal basis from $\left(\begin{bmatrix} 1 \\ 0 \end{bmatrix}, \begin{bmatrix} 0 \\ 1 \end{bmatrix}\right)$ to $\left(\begin{bmatrix} \cos(\alpha) \\ -\sin(\alpha) \end{bmatrix}, \begin{bmatrix} \sin(\alpha) \\ \cos(\alpha) \end{bmatrix}\right)$.

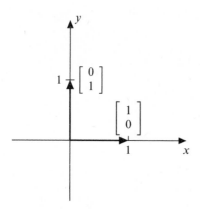

Figure 3.1
The standard basis.

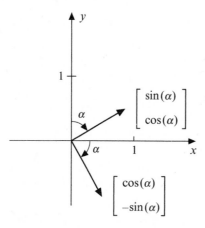

Figure 3.2
The standard basis rotated by $\alpha°$.

If the basis is rotated through 90° it becomes $\left(\begin{bmatrix} \cos(90°) \\ -\sin(90°) \end{bmatrix}, \begin{bmatrix} \sin(90°) \\ \cos(90°) \end{bmatrix}\right)$, which simplifies to $\left(\begin{bmatrix} 0 \\ -1 \end{bmatrix}, \begin{bmatrix} 1 \\ 0 \end{bmatrix}\right)$. As we previously noted, $\begin{bmatrix} 0 \\ -1 \end{bmatrix}$ is equivalent to $\begin{bmatrix} 0 \\ 1 \end{bmatrix}$, so rotating through 90° brings us back to a basis equivalent to the original one, except that the order of the basis elements has been interchanged (i.e., N and S have been interchanged).

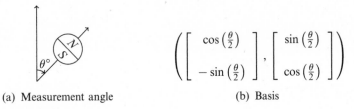

(a) Measurement angle   (b) Basis

Figure 3.3
Rotating measuring apparatus by $\theta°$.

We will let θ denote the angle we are rotating our measurement apparatus and α the angle we rotate our basis vectors. We have seen that we
get a complete set of directions as θ goes from $0°$ to $180°$, and that we get
a complete set of rotated bases as α goes from $0°$ to $90°$. Once we reach
$\theta = 180°$ or equivalently $\alpha = 90°$, N and S measured in direction $0°$ are
interchanged.

We make the natural definition that $\theta = 2\alpha$. Consequently, the basis
associated with rotating our apparatus by θ is $\left(\begin{bmatrix} \cos(\theta/2) \\ -\sin(\theta/2) \end{bmatrix}, \begin{bmatrix} \sin(\theta/2) \\ \cos(\theta/2) \end{bmatrix} \right)$.
Figure 3.3 illustrates this.

Rotating the Apparatus through 60°

As an example to illustrate our formula, we look at what happens when
we rotate our measuring apparatus by $60°$. Suppose that we first measured
our electron to have spin N in direction $0°$. We will measure it again using
the apparatus turned through $60°$. What is the probability that it gives
a result of N?

In this case the associated basis to the rotated apparatus is
$\left(\begin{bmatrix} \cos(30°) \\ -\sin(30°) \end{bmatrix}, \begin{bmatrix} \sin(30°) \\ \cos(30°) \end{bmatrix} \right)$ which simplifies to $\left(\begin{bmatrix} \sqrt{3}/2 \\ -1/2 \end{bmatrix}, \begin{bmatrix} 1/2 \\ \sqrt{3}/2 \end{bmatrix} \right)$.

Since the electron initially was measured to have spin N in direction $0°$,
its state vector after the initial measurement was $\begin{bmatrix} 1 \\ 0 \end{bmatrix}$. We must now express
this as a linear combination of the new basis vectors. To get the coordinates
relative to the new basis we can multiply the state vector on the left by the
matrix consisting of the bras of the basis. This gives:

$$\begin{bmatrix} \sqrt{3}/2 & -1/2 \\ 1/2 & \sqrt{3}/2 \end{bmatrix}\begin{bmatrix} 1 \\ 0 \end{bmatrix} = \begin{bmatrix} \sqrt{3}/2 \\ 1/2 \end{bmatrix},$$

telling us that

$$\begin{bmatrix} 1 \\ 0 \end{bmatrix} = \sqrt{3}/2 \begin{bmatrix} \sqrt{3}/2 \\ -1/2 \end{bmatrix} + 1/2 \begin{bmatrix} 1/2 \\ \sqrt{3}/2 \end{bmatrix}.$$

So the probability of getting N when we measure in the 60° direction is $\left(\sqrt{3}\big/2\right)^2 = 3/4$.

The Mathematical Model for Photon Polarization

In most of the book we will restrict our attention to measuring spin of electrons, but in the first chapter we said that we could rewrite everything in terms of the polarization of photons. In the next few sections we will explain the analogy between electron spin and photon polarization and give the mathematical model of polarization.

We start by associating the angle of 0° with a polarized filter in the vertical direction, that is, a filter that lets through photons that are polarized vertically, which means that horizontally polarized photons are absorbed by the filter. As with the spin of electrons, we associate the standard basis $\left(\begin{bmatrix} 1 \\ 0 \end{bmatrix}, \begin{bmatrix} 0 \\ 1 \end{bmatrix}\right)$ to the angle of 0°. The vector $\begin{bmatrix} 1 \\ 0 \end{bmatrix}$ corresponds to a vertically polarized photon and the vector $\begin{bmatrix} 0 \\ 1 \end{bmatrix}$ to a horizontally polarized one.

We will rotate the filter through an angle $\beta°$. It now lets through photons that are polarized in direction $\beta°$ and blocks photons that are polarized perpendicularly to $\beta°$.

The mathematical model follows that for the spin of electrons. For each direction, there is an ordered orthonormal basis $(|b_1\rangle, |b_2\rangle)$ associated with making a polarization measurement in this direction. The ket $|b_1\rangle$ corresponds to a photon that is polarized in the given direction—that is, that passes through the filter. The ket $|b_2\rangle$ corresponds to a photon that is polarized orthogonally to the given direction—that is absorbed by the filter.

A photon has a polarization state given by a ket, $|v\rangle$. This can be written as a linear combination of the vectors in the basis: $|v\rangle = d_1 |b_1\rangle + d_2 |b_2\rangle$.

When the polarization is measured in the direction given by the ordered basis, the result will be that the photon is polarized in the given direction with probability d_1^2 and polarized perpendicularly with probability d_2^2; that is, the probability the photon passes through the filter is d_1^2, and the probability it is absorbed is d_2^2.

If the result of the measurement is that the photon is polarized in the given direction—it passes through the filter—then the state of the photon becomes $|b_1\rangle$.

The Basis Associated with a Given Polarization Direction

Recall that if we start with our standard basis $\left(\begin{bmatrix}1\\0\end{bmatrix},\begin{bmatrix}0\\1\end{bmatrix}\right)$ and rotate these vectors though an angle α, we obtain the new orthonormal basis $\left(\begin{bmatrix}\cos(\alpha)\\-\sin(\alpha)\end{bmatrix},\begin{bmatrix}\sin(\alpha)\\\cos(\alpha)\end{bmatrix}\right)$. Also recall that rotating through an angle of $90°$ brings us back to the same basis as the original, except that the order of the basis elements has been interchanged.

Now consider rotating a polarized filter through an angle β. When β is $0°$, we are measuring in the vertical and horizontal direction. The vertically polarized photons pass through the filter, and the horizontally polarized photons are absorbed. Once β reaches $90°$ we will be measuring photons in the vertical and horizontal direction, but now the vertically polarized photons are absorbed and the horizontally polarized ones pass through. In this case, $\beta = 90°$ corresponds to $\alpha = 90°$, and, in general, we can take $\alpha = \beta$.

In conclusion, the ordered orthonormal basis associated with rotating a polarized filter through an angle β is $\left(\begin{bmatrix}\cos(\beta)\\-\sin(\beta)\end{bmatrix},\begin{bmatrix}\sin(\beta)\\\cos(\beta)\end{bmatrix}\right)$.

The Polarized Filters Experiments

Using our model, we describe the experiments that we looked at in the first chapter.

In the first experiment we have two polarized squares. One measures polarization in direction $0°$ and the other in direction $90°$. No light is let through the region of overlap, as depicted in figure 3.4.

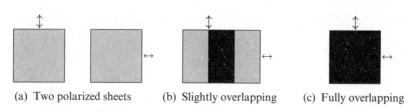

(a) Two polarized sheets   (b) Slightly overlapping   (c) Fully overlapping

Figure 3.4
Two polarized squares.

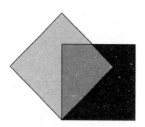

Figure 3.5
Three polarized squares.

The basis associated with 0° is the standard orthonormal basis. The basis associated with 90° is the same, except the order of the elements has been changed. A photon that passes through the first filter has had a measurement made—it is vertically polarized—and so is now in state $\begin{bmatrix} 1 \\ 0 \end{bmatrix}$. We now measure it with the second filter. This lets through photons with state vector $\begin{bmatrix} 0 \\ 1 \end{bmatrix}$ and absorbs photons with state vector $\begin{bmatrix} 1 \\ 0 \end{bmatrix}$. Consequently, any photon that passes through the first filter is absorbed by the second.

In the three-filter experiment we have the two filters arranged as above. We take the third sheet and rotate it through 45°, and slide this sheet between the other two. Some light comes through the region of overlap of all three squares. This is depicted in figure 3.5.

The ordered bases for the three filters are $\left(\begin{bmatrix} 1 \\ 0 \end{bmatrix}, \begin{bmatrix} 0 \\ 1 \end{bmatrix} \right)$, $\left(\begin{bmatrix} \frac{1}{\sqrt{2}} \\ \frac{-1}{\sqrt{2}} \end{bmatrix}, \begin{bmatrix} \frac{1}{\sqrt{2}} \\ \frac{1}{\sqrt{2}} \end{bmatrix} \right)$ and $\left(\begin{bmatrix} 0 \\ 1 \end{bmatrix}, \begin{bmatrix} 1 \\ 0 \end{bmatrix} \right)$. A photon that passes through all three filters will have had three measurements made. Photons that pass through the first filter will be in state $\begin{bmatrix} 1 \\ 0 \end{bmatrix}$.

The second measurement corresponds to passing through the filter rotated by 45°. We need to rewrite the state of the photon using the appropriate basis.

$$\begin{bmatrix} 1 \\ 0 \end{bmatrix} = \frac{1}{\sqrt{2}} \begin{bmatrix} \frac{1}{\sqrt{2}} \\ \frac{-1}{\sqrt{2}} \end{bmatrix} + \frac{1}{\sqrt{2}} \begin{bmatrix} \frac{1}{\sqrt{2}} \\ \frac{1}{\sqrt{2}} \end{bmatrix}$$

The probability of a photon passing through the second filter once it has gone through the first is $\left(\frac{1}{\sqrt{2}} \right)^2 = \frac{1}{2}$. Consequently, half the photons that pass through the first filter will pass through the second filter. Those that do will now be in state $\begin{bmatrix} \frac{1}{\sqrt{2}} \\ \frac{-1}{\sqrt{2}} \end{bmatrix}$.

The third filter corresponds to making a measurement using the third basis. We must rewrite the state of our photon using this basis.

$$\begin{bmatrix} \frac{1}{\sqrt{2}} \\ \frac{-1}{\sqrt{2}} \end{bmatrix} = \frac{-1}{\sqrt{2}} \begin{bmatrix} 0 \\ 1 \end{bmatrix} + \frac{1}{\sqrt{2}} \begin{bmatrix} 1 \\ 0 \end{bmatrix}$$

The third filter lets through photons corresponding to state $\begin{bmatrix} 0 \\ 1 \end{bmatrix}$. The probability of this is $\left(\frac{-1}{\sqrt{2}} \right)^2 = \frac{1}{2}$. Consequently, half the photons that pass through the first two filters will pass through the third filter.

We have shown how the mathematical model relates the spin of an electron to the polarization of a photon. This model is also exactly what we need to describe qubits.

Qubits

A classical bit is either 0 or 1. It can be represented by anything that has two mutually exclusive states. The standard example is a switch that can be in either the on or off position. In classical computer science the measurement of bits does not enter the picture. A bit is a bit. It is either 0 or it is 1, and that is all there is to it. But for qubits the situation is more complicated, and measurement is a crucial part of the mathematical description.

We define a *qubit* to be any unit ket in $\mathbb{R}^2$. Usually, given a qubit, we will want to measure it. If we are going to measure it, we also need to include a direction of measurement. This is done by introducing an ordered orthonormal basis $(|b_0\rangle, |b_1\rangle)$. The qubit can be written as a linear combination— often called a linear superposition—of the basis vectors. In general, it will have the form $d_0 |b_0\rangle + d_1 |b_1\rangle$. After we measure, its state will jump to either $|b_0\rangle$ or $|b_1\rangle$. The probability of its being $|b_0\rangle$ is d_0^2; the probability of $|b_1\rangle$ is d_1^2. This is exactly the same model we have been using, but now we connect the classical bits 0 and 1 to the basis vectors. We will associate the $|b_0\rangle$ basis vector with the bit 0 and the $|b_1\rangle$ basis vector with the bit 1. So when we measure the qubit $d_0 |b_0\rangle + d_1 |b_1\rangle$ we will obtain 0 with probability d_0^2 and 1 with probability d_1^2.

Since a qubit can be any unit ket and there are infinitely many unit kets, there are infinitely many possible values for a qubit. This is quite unlike classical computation, where we just have two bits. It is important, however, to notice that to get information out of a qubit we have to measure it. When we measure it we will get either 0 or 1, so the result is a classical bit.

We will give some illustrative examples using Alice, Bob, and Eve.

Alice, Bob, and Eve

Alice, Bob, and Eve are three characters that often appear in cryptography. Alice wants to send a confidential message to Bob. Unfortunately, Eve wants to eavesdrop with evil intent. How should Alice encrypt her messages so that Bob can read them but Eve cannot? This is the central question of cryptography. We will look at it later. But for the moment we will just concentrate on Alice sending Bob a stream of qubits.

Alice measures qubits using her orthonormal basis, which we will denote as $(|a_0\rangle, |a_1\rangle)$. Bob measures the qubits that Alice sends to him using his orthonormal basis $(|b_0\rangle, |b_1\rangle)$.

Suppose that Alice wants to send 0. She can use her measuring apparatus to sort qubits into either state $|a_0\rangle$ or $|a_1\rangle$. Since she wants to send 0, she sends a qubit in state $|a_0\rangle$. Bob is measuring with respect to his ordered basis. To calculate what happens we must write $|a_0\rangle$ as a linear combination of Bob's basis vectors. It will have the form $|a_0\rangle = d_0|b_0\rangle + d_1|b_1\rangle$. When Bob measures the qubit one of two things happens: Either it jumps to state $|b_0\rangle$ with probability d_0^2 and he writes down 0, or it jumps to state $|b_1\rangle$ with probability d_1^2 and he writes down 1.

You might be wondering why Bob and Alice don't choose to use the same basis. If they did this, Bob would receive 0 with certainty whenever Alice sent 0 and receive 1 with certainty whenever Alice sent 1. This is true, but remember Eve. If she also chooses the same basis, then she too will receive exactly the same message as Bob. We will see later that there are good reasons for why Alice and Bob might choose different bases to thwart Eve.

For an example, Alice and Bob might choose to measure their qubits using either the basis $\left(\begin{bmatrix}1\\0\end{bmatrix}, \begin{bmatrix}0\\1\end{bmatrix}\right)$ or the basis $\left(\begin{bmatrix}\frac{1}{\sqrt{2}}\\\frac{-1}{\sqrt{2}}\end{bmatrix}, \begin{bmatrix}\frac{1}{\sqrt{2}}\\\frac{1}{\sqrt{2}}\end{bmatrix}\right)$. The calculations are exactly as before, where we were considering spin in the vertical and horizontal directions. The only change is that we replace N with 0 and S with 1. Only if Alice and Bob choose to use the same basis will Bob end up with exactly the bit that Alice wanted to send. If they choose to use different bases, then half of the time Bob gets the correct bit, but half of the time he gets the wrong one. This might not seem very useful, but we will see at the end of this chapter that Alice and Bob can use these two bases to secure their communications.

A couple of chapters from now, Alice and Bob will each choose one of three bases at random. These correspond to measuring the spin of an electron in the directions of 0°, 120°, or 240°. We will need to analyze all the possibilities, but now, to give a concrete example, we will have Alice measure in the 240° direction and Bob in the 120° direction.

We know the orthonormal basis in direction θ is $\left(\begin{bmatrix}\cos(\theta/2)\\-\sin(\theta/2)\end{bmatrix},\right.$ $\left.\begin{bmatrix}\sin(\theta/2)\\\cos(\theta/2)\end{bmatrix}\right)$. Consequently, Alice's basis is $\left(\begin{bmatrix}-1/2\\-\sqrt{3}/2\end{bmatrix}, \begin{bmatrix}\sqrt{3}/2\\-1/2\end{bmatrix}\right)$ and Bob's is $\left(\begin{bmatrix}1/2\\-\sqrt{3}/2\end{bmatrix}, \begin{bmatrix}\sqrt{3}/2\\1/2\end{bmatrix}\right)$. Since multiplying a ket by -1 gives an equivalent ket, we can simplify Alice's basis to $\left(\begin{bmatrix}1/2\\\sqrt{3}/2\end{bmatrix}, \begin{bmatrix}\sqrt{3}/2\\-1/2\end{bmatrix}\right)$. (Notice that this is the basis for direction 60° that we looked at earlier, with the order of the basis vectors switched. There is nothing surprising about this. In fact, that's exactly what we expect. Measuring N in direction 240° is exactly the same as measuring S in direction 60°.)

If Alice wants to send 0, she sends the qubit $\begin{bmatrix} 1/2 \\ \sqrt{3}/2 \end{bmatrix}$. To calculate what Bob measures we need to write this as a linear superposition of his basis vectors. We can get the probability amplitudes by forming the matrix consisting of the bras of his basis vectors and then multiplying the qubit by this matrix.

$$\begin{bmatrix} 1/2 & -\sqrt{3}/2 \\ \sqrt{3}/2 & 1/2 \end{bmatrix}\begin{bmatrix} 1/2 \\ \sqrt{3}/2 \end{bmatrix} = \begin{bmatrix} -1/2 \\ \sqrt{3}/2 \end{bmatrix}.$$

This tells us that

$$\begin{bmatrix} 1/2 \\ \sqrt{3}/2 \end{bmatrix} = -1/2\begin{bmatrix} 1/2 \\ -\sqrt{3}/2 \end{bmatrix} + \sqrt{3}/2\begin{bmatrix} \sqrt{3}/2 \\ 1/2 \end{bmatrix}.$$

This means that when Bob measures the qubit, he gets 0 with probability 1/4 and 1 with probability 3/4. Similarly, it can be checked that if Alice sends 1, Bob will get 1 with probability 1/4 and 0 with probability 3/4.

It can also be checked, and it is an excellent exercise, that if Alice and Bob choose from the three bases, where the third is the standard basis, and end up with different bases, then Bob always gets the correct bit with probability 1/4.

Probability Amplitudes and Interference

If you drop a stone into a pond, waves propagate outward from where the stone hits the water. If you drop two stones, the waves propagating form one stone can interfere with the waves coming from the other one. If the waves are in phase—the peaks or the troughs coincide—then you get constructive interference: The amplitude of the resulting wave increases. If the waves are out of phase—the peak of one meets the trough of the other—then you get destructive interference: The amplitude of the resulting wave decreases.

A qubit has the form $d_0|b_0\rangle + d_1|b_1\rangle$, where d_0 and d_1 are the probability amplitudes. The square of these numbers gives the probabilities that the qubit jumps to the corresponding basis vector. Probabilities are not allowed to be negative, but probability amplitudes can be. This fact allows both constructive and destructive interference to take place.

As an example, consider the qubits that we are denoting by $|\leftarrow\rangle$ and $|\rightarrow\rangle$ If we measure either of them in the standard basis, they will jump to either $|\uparrow\rangle$ or $|\downarrow\rangle$. Each has a probability of 1/2 of occurring. If we are translating this back to bits, we will get either a 0 or a 1 with equal probability. We now take a superposition of the original two qubits, $|v\rangle = \frac{1}{\sqrt{2}}|\leftarrow\rangle + \frac{1}{\sqrt{2}}|\rightarrow\rangle$. If we were to measure $|v\rangle$ in the horizontal direction, we would get either $|\leftarrow\rangle$ or $|\rightarrow\rangle$ with equal probability. But if we measure in the vertical direction we get 0 with certainty, because

$$|v\rangle = \frac{1}{\sqrt{2}}|\leftarrow\rangle + \frac{1}{\sqrt{2}}|\rightarrow\rangle = \frac{1}{\sqrt{2}}\begin{bmatrix} \frac{1}{\sqrt{2}} \\ \frac{1}{\sqrt{2}} \end{bmatrix} + \frac{1}{\sqrt{2}}\begin{bmatrix} \frac{1}{\sqrt{2}} \\ \frac{-1}{\sqrt{2}} \end{bmatrix} = 1\begin{bmatrix} 1 \\ 0 \end{bmatrix} + 0\begin{bmatrix} 0 \\ 1 \end{bmatrix}.$$

The terms in $|\leftarrow\rangle$ and $|\rightarrow\rangle$ that give 0 have interfered constructively, and the terms that give 1 have interfered destructively.

This will be important when it comes to quantum algorithms. We want to choose linear combinations carefully so that terms that we are not interested in cancel, but terms that we are interested in are amplified.

There are a very limited number of things that we can do with one qubit, but one thing we can do is to enable Alice and Bob to communicate securely.

Alice, Bob, Eve, and the BB84 Protocol

We often want to send secure messages. All Internet commerce depends on it. The standard way that messages are encrypted and decrypted uses two steps. The first step is when first contact is made. The two parties agree on a key—a long string of binary digits. Once they both have the same key, they then use it to both encode and decode messages from one another. The security of the method comes from the key. It is impossible to decode the messages between the two parties without knowing the key.

Alice and Bob want to communicate securely. Eve wants to eavesdrop. Alice and Bob want to agree on a key, but they need to be sure that Eve does not know it.

The BB84 protocol derives its name from its inventors, Charles Bennett and Gilles Brassard, and the year that it was invented, 1984. It uses two sets of ordered, orthonormal bases: the standard one, $\left(\begin{bmatrix} 1 \\ 0 \end{bmatrix}, \begin{bmatrix} 0 \\ 1 \end{bmatrix} \right)$, that we used for measuring spin in the vertical direction, and so is denoted by V, and $\left(\begin{bmatrix} \frac{1}{\sqrt{2}} \\ \frac{-1}{\sqrt{2}} \end{bmatrix}, \begin{bmatrix} \frac{1}{\sqrt{2}} \\ \frac{1}{\sqrt{2}} \end{bmatrix} \right)$ that we used for measuring spin in the horizontal direction, and so is denoted by H. In both cases, the classical bit 0 will correspond to the first vector in the ordered basis and 1 to the second.

Alice chooses the key that she wants to send to Bob. This is a string of classical bits. For each bit, Alice chooses one of the two bases V and H at random and with equal probability. She then sends Bob the qubit consisting of the appropriate basis vector. For example, if she wants to send 0 and chooses V, she will send $\begin{bmatrix} 1 \\ 0 \end{bmatrix}$, if she chooses H, she will send $\begin{bmatrix} \frac{1}{\sqrt{2}} \\ \frac{-1}{\sqrt{2}} \end{bmatrix}$. She follows the same process for each bit, keeping a record of which basis she has used for each bit. If the string is $4n$ binary digits long, she will end up with a string of length $4n$ consisting of Vs and Hs. (The reason we are using $4n$ and not n will become clear in a moment, but n should be a fairly large number.)

Bob also chooses between the two bases at random and with equal probability. He then measures the qubit in his chosen basis. Bob does this for each bit, and he keeps a record of which basis he has used. At the end of the transmission he also ends up with two strings of length $4n$, one consisting of 0s and 1s from his measurements, the other consisting of Vs and Hs corresponding to the bases he chose.

Alice and Bob are choosing the basis for each bit at random. Half the time they end up using the same basis, while half the time they use different bases. If they both choose the same basis, then Bob will obtain the bit that Alice is sending with certainty. If they choose different bases, then half the time Bob gets the right bit, but half the time it is the wrong bit—no information is transmitted when they choose different bases.

Alice and Bob now compare their strings of Vs and Hs over an unencrypted line. They keep the bits corresponding to the times when they both used the same basis and erase the bits that correspond to times that they used different bases. If Eve is not intercepting the message, they both end up with the same string of binary digits that has length about $2n$. They now must check to see if Eve was listening in.

If Eve intercepts the qubit on the way from Alice to Bob, she would really like to clone it, sending one copy on to Bob and measuring the other qubit. Unfortunately for Eve, this is impossible. To obtain any information, she has to measure the qubit that Alice has sent, and this could change the qubit—it will end up as one of the basis vectors in the basis with which she chooses to measure. The best she can do is to choose one of the two bases at random, measure the qubit, and then send the qubit on to Bob. Let's see what happens.

Alice and Bob are interested only in the measurements where they chose the same basis. We will restrict our attention to these times. When Alice and Bob agree on the basis, half the time Eve will also agree, and half the time she will choose the other basis. If all three agree on the basis, then they will all get the same bit as the measurement. If Eve chooses the wrong basis, then she will send a qubit that is in a superposition of Bob's basis states. When Bob measures this qubit he will get 0 and 1 with equal probability; he will get the right bit one half of the time.

We now return to Alice and Bob and their strings of bits of length, at the moment, of $2n$. They know that if Eve is not intercepting qubits, these strings will be identical. But they know that if Eve is intercepting qubits, she is going to choose the wrong basis half the time, and in these cases Bob will end up with the wrong bit half of the time. So, if Eve is intercepting qubits, a quarter of Bob's bits will disagree with Alice's. They now compare half of the $2n$ bits over an unencrypted line. If they agree on all of them, they know Eve is not listening in and can use the other n bits as the key. If they disagree on a quarter of the bits, they know that Eve is intercepting their qubits. They know that they need to find another way to secure their communication.

This is a nice example of sending one qubit at a time. There are, however, very few things that we can do with qubits that don't interact with one another. In the next chapter we look at what happens when we have two or more qubits. In particular, we look at yet another phenomenon that is not part of our classical worldview but that plays an essential part of the quantum world: entanglement.

4 Entanglement

In this chapter we study the mathematics of entanglement. To do this, we need to introduce one more idea from linear algebra: the tensor product. We start by looking at two systems with no interaction between them. Since there is no interaction, we could study each system by itself, without any reference to the other system, but we will show how we can combine the two systems using tensor products. Then we introduce the tensor product of two vector spaces and show that most of the vectors in this product represent what are referred to as entangled states.

Throughout this chapter there will be two qubits. Alice has one, and Bob has another. We will begin our study by examining a case where there is no interaction between Alice's and Bob's systems. This analysis initially might seem to make something that is very simple look somewhat complicated, but once we have described everything in terms of tensor products it becomes fairly straightforward to extend the underlying ideas to the general entangled case.

The approach we take, however, is not the approach we have taken so far. Instead of presenting physical experiments and then deriving a mathematical model, we proceed in the other direction. We will extend our model in the simplest way possible and then see what the model predicts should be found when we perform experiments. We find that the model predicts the experiments accurately, but the conclusions are quite surprising.

Alice and Bob's Qubits Are Not Entangled

We suppose that Alice is measuring using the orthonormal basis $(|a_0\rangle, |a_1\rangle)$ and Bob is measuring with orthonormal basis $(|b_0\rangle, |b_1\rangle)$. A typical qubit for Alice is $|v\rangle = c_0|a_0\rangle + c_1|a_1\rangle$, and for Bob is $|w\rangle = d_0|b_0\rangle + d_1|b_1\rangle$. We can combine these two state vectors using a new type of product that we call a *tensor* product, giving us a new vector denoted by $|v\rangle \otimes |w\rangle$.

Now $|v\rangle \otimes |w\rangle = (c_0 |a_0\rangle + c_1 |a_1\rangle) \otimes (d_0 |b_0\rangle + d_1 |b_1\rangle)$. How do we multiply these two terms using this new product? Well, we do it the most natural way possible. We expand it in the usual way we multiply out algebraic expressions of the form $(a + b)(c + d)$. We write

$$(c_0 |a_0\rangle + c_1 |a_1\rangle) \otimes (d_0 |b_0\rangle + d_1 |b_1\rangle)$$
$$= c_0 d_0 |a_0\rangle \otimes |b_0\rangle + c_0 d_1 |a_0\rangle \otimes |b_1\rangle + c_1 d_0 |a_1\rangle \otimes |b_0\rangle + c_1 d_1 |a_1\rangle \otimes |b_1\rangle$$

If you are familiar with the *FOIL* method, you should recognize that this is exactly what we have done. To make the terminology even simpler we will use juxtaposition of two kets to mean the tensor product, so $|v\rangle \otimes |w\rangle$ will be denoted as $|v\rangle |w\rangle$.

$$|v\rangle |w\rangle = (c_0 |a_0\rangle + c_1 |a_1\rangle)(d_0 |b_0\rangle + d_1 |b_1\rangle)$$
$$= c_0 d_0 |a_0\rangle |b_0\rangle + c_0 d_1 |a_0\rangle |b_1\rangle + c_1 d_0 |a_1\rangle |b_0\rangle + c_1 d_1 |a_1\rangle |b_1\rangle$$

Though this is just the standard way of multiplying out two expressions, there is one thing that we have to be very careful about: The first ket in the tensor product belongs to Alice, and the second ket belongs to Bob. For example, $|v\rangle |w\rangle$ means that $|v\rangle$ belongs to Alice and that $|w\rangle$ belongs to Bob. The product $|w\rangle |v\rangle$ means that $|w\rangle$ belongs to Alice and that $|v\rangle$ belongs to Bob. So, in general, $|v\rangle |w\rangle$ will not equal $|w\rangle |v\rangle$. The technical term for this is to say the tensor product is not commutative.

Alice is measuring with her orthonormal basis ($|a_0\rangle, |a_1\rangle$) and Bob is measuring with his orthonormal basis ($|b_0\rangle, |b_1\rangle$). We are describing both Alice's and Bob's qubits using tensor notation. This description involves the four tensor products that come from the basis vectors: $|a_0\rangle |b_0\rangle$, $|a_0\rangle |b_1\rangle$, $|a_1\rangle |b_0\rangle$, and $|a_1\rangle |b_1\rangle$. These four products form an orthonormal basis for the tensor product of Alice's and Bob's systems: Each of these products is a unit vector, and they are orthogonal to one another.

At this stage, though we have introduced new notation, we have not introduced anything new in terms of concepts. It is just the information that we already knew, but in a different package. For example, the number $c_0 d_0$ is a probability amplitude. Its square gives the probability that when both Alice and Bob measure their qubits Alice's qubit jumps to $|a_0\rangle$, that is, she reads 0, and Bob's qubit jumps to $|b_0\rangle$, that is, he reads 0. But we already knew that the probability that Alice's qubit would jump to $|a_0\rangle$ is c_0^2 and the probability that Bob's would jump to $|b_0\rangle$ is d_0^2. So we really knew that the probability that both would occur is $c_0^2 d_0^2$, which is, of course, the same as $(c_0 d_0)^2$. In a similar way, the numbers $c_0^2 d_1^2$, $c_1^2 d_0^2$, and $c_1^2 d_1^2$ give the probabilities that Alice and Bob read 01, 10, and 11, respectively. (Remember that Alice's bit is always listed before Bob's.)

Next we will replace these probability amplitudes using just one symbol instead of two. We will let $r = c_0 d_0$, $s = c_0 d_1$, $t = c_1 d_0$, and $u = c_1 d_1$, so $|v\rangle|w\rangle = r|a_0\rangle|b_0\rangle + s|a_0\rangle|b_1\rangle + t|a_1\rangle|b_0\rangle + u|a_1\rangle|b_1\rangle$. We know that $r^2 + s^2 + t^2 + u^2 = 1$, because they are probability amplitudes. We also know that $ru = st$, because both ru and st equal $c_0 c_1 d_0 d_1$. Now we come to the new idea. We are going to describe the states of Alice's and Bob's qubits by tensors of the form $r|a_0\rangle|b_0\rangle + s|a_0\rangle|b_1\rangle + t|a_1\rangle|b_0\rangle + u|a_1\rangle|b_1\rangle$. Again we stipulate that $r^2 + s^2 + t^2 + u^2 = 1$, so that we can treat r, s, t, and u as probability amplitudes. But we no longer insist that $ru = st$. We allow any values of r, s, t, and u just as long as the sum of their squares is 1.

Given a tensor of the form $r|a_0\rangle|b_0\rangle + s|a_0\rangle|b_1\rangle + t|a_1\rangle|b_0\rangle + u|a_1\rangle|b_1\rangle$ with $r^2 + s^2 + t^2 + u^2 = 1$, there are two cases. The first case is when $ru = st$. In this case we say that Alice's and Bob's qubits are *not entangled*. The second case is when $ru \neq st$. In this case we say that Alice's and Bob's qubits are *entangled*. This rule is easy to remember if the terms are written out with the subscripts in the order we have presented them: 00, 01, 10, 11. In this order, ru are the outer terms and st are the inner, so the qubits are not entangled if the product of the outer terms is equal to the product of the inner ones, and they are entangled if the products are not equal.

We will look at examples that illustrate both of these cases.

Unentangled Qubits Calculation

Suppose we are told that Alice and Bob's qubits are given by

$$\frac{1}{2\sqrt{2}}|a_0\rangle|b_0\rangle + \frac{\sqrt{3}}{2\sqrt{2}}|a_0\rangle|b_1\rangle + \frac{1}{2\sqrt{2}}|a_1\rangle|b_0\rangle + \frac{\sqrt{3}}{2\sqrt{2}}|a_1\rangle|b_1\rangle.$$

We quickly calculate the products of the outer and inner probability amplitudes. Both products equal $\sqrt{3}/8$, so the qubits are unentangled.

The probability amplitudes tell us what happens when Alice and Bob both make measurements. They will get 00 with probability 1/8, 01 with probability 3/8, 10 with probability 1/8 and 11 with probability 3/8.

What is slightly trickier is to see what happens if just one of them makes a measurement. We start by assuming that Alice is going to make a measurement, but Bob is not. To do this we begin by pulling out common factors from Alice's perspective. We rewrite the tensor product as

$$|a_0\rangle\left(\frac{1}{2\sqrt{2}}|b_0\rangle + \frac{\sqrt{3}}{2\sqrt{2}}|b_1\rangle\right) + |a_1\rangle\left(\frac{1}{2\sqrt{2}}|b_0\rangle + \frac{\sqrt{3}}{2\sqrt{2}}|b_1\rangle\right).$$

Next, we want the expressions in the parentheses to be unit vectors, so we divide by their lengths inside the parentheses and multiply by their lengths outside, which gives us

$$\frac{1}{\sqrt{2}}|a_0\rangle\left(\frac{1}{2}|b_0\rangle + \frac{\sqrt{3}}{2}|b_1\rangle\right) + \frac{1}{\sqrt{2}}|a_1\rangle\left(\frac{1}{2}|b_0\rangle + \frac{\sqrt{3}}{2}|b_1\rangle\right).$$

We can then pull out the common factor in the parentheses. (But remember that it's Bob's, so we must keep it on the right.)

$$\left(\frac{1}{\sqrt{2}}|a_0\rangle + \frac{1}{\sqrt{2}}|a_1\rangle\right)\left(\frac{1}{2}|b_0\rangle + \frac{\sqrt{3}}{2}|b_1\rangle\right).$$

Written in this form, it becomes explicit that the states are not entangled. We have a tensor product of a qubit that belongs to Alice with a qubit that belongs to Bob.

From this we can deduce that if Alice measures first she will obtain 0 and 1 with equal probability. This measurement has no effect on the state of Bob's qubit. It was and remains

$$\left(\frac{1}{2}|b_0\rangle + \frac{\sqrt{3}}{2}|b_1\rangle\right).$$

We can also read from the factored expression that if Bob measures first he will get 0 with probability 1/4 and 1 with probability 3/4. Again, it is clear that Bob's measurement has no effect on Alice's qubit.

When the qubits are unentangled, a measurement of one of the qubits has absolutely no effect on the other qubit. The situation is completely different with entangled qubits. If qubits are entangled, the measurement of one will have an effect on the other one. We will illustrate this with an example.

Entangled Qubits Calculation

Suppose we are told that Alice's and Bob's qubits are given by

$$\frac{1}{2}|a_0\rangle|b_0\rangle + \frac{1}{2}|a_0\rangle|b_1\rangle + \frac{1}{\sqrt{2}}|a_1\rangle|b_0\rangle + 0|a_1\rangle|b_1\rangle.$$

We quickly calculate the products of the outer and inner probability. The product of the outer terms is 0. Since the product of the inner terms is not equal to 0, the two qubits are entangled.

Usually both Alice and Bob will make measurements. As in the previous example, we can use the probability amplitudes to tell us what happens when Alice and Bob both measure their qubits. They will get 00 with probability 1/4, 01 with probability 1/4, 10 with probability 1/2, and 11 with probability 0. Notice that there is nothing strange going on. This is exactly the same calculation as in the unentangled case.

We will now see what happens if just one of them makes a measurement. We start by assuming that Alice is going to make a measurement, but Bob is not. To do this we begin by pulling out common factors from Alice's perspective. We rewrite the tensor product as

$$|a_0\rangle\left(\frac{1}{2}|b_0\rangle + \frac{1}{2}|b_1\rangle\right) + |a_1\rangle\left(\frac{1}{\sqrt{2}}|b_0\rangle + 0|b_1\rangle\right).$$

As before, we want the expressions in the parentheses to be unit vectors, so we divide by their lengths inside the parentheses and multiply by their lengths outside, which gives us

$$\frac{1}{\sqrt{2}}|a_0\rangle\left(\frac{1}{\sqrt{2}}|b_0\rangle + \frac{1}{\sqrt{2}}|b_1\rangle\right) + \frac{1}{\sqrt{2}}|a_1\rangle(1|b_0\rangle + 0|b_1\rangle).$$

In the previous example, the terms in the parentheses were the same, and we could pull this common term out as a common factor. But in this case the terms in the parentheses are different. This is what it means to be entangled.

The probability amplitudes in front of Alice's kets tell us that when she measures she will get 0 and 1 with equal probability. But when Alice gets 0, her qubit jumps to $|a_0\rangle$. The combined system jumps to the unentangled state $|a_0\rangle\left(\frac{1}{\sqrt{2}}|b_0\rangle + \frac{1}{\sqrt{2}}|b_1\rangle\right)$, and Bob's qubit is no longer entangled with Alice's. It is $\left(\frac{1}{\sqrt{2}}|b_0\rangle + \frac{1}{\sqrt{2}}|b_1\rangle\right)$. When Alice gets 1, again Bob's qubit is no longer entangled with Alice's. It becomes $|b_0\rangle$.

The result of Alice's measurement affects Bob's qubit. If she gets 0, Bob's qubit becomes $\left(\frac{1}{\sqrt{2}}|b_0\rangle + \frac{1}{\sqrt{2}}|b_1\rangle\right)$. If she gets 1, his qubit becomes $|b_0\rangle$. This does seem strange. Alice and Bob could be far apart. As soon as she makes a measurement Bob's qubit becomes unentangled, but exactly what it is depends on Alice's outcome.

For completeness, we will see what happens when Bob measures first.

We start with the initial tensor product.

$$\frac{1}{2}|a_0\rangle|b_0\rangle + \frac{1}{2}|a_0\rangle|b_1\rangle + \frac{1}{\sqrt{2}}|a_1\rangle|b_0\rangle + 0|a_1\rangle|b_1\rangle$$

Rewriting from Bob's perspective gives

$$\left(\frac{1}{2}|a_0\rangle + \frac{1}{\sqrt{2}}|a_1\rangle\right)|b_0\rangle + \left(\frac{1}{2}|a_0\rangle + 0|a_1\rangle\right)|b_1\rangle.$$

As always, we want the expressions in the parentheses to be unit vectors, so we divide by their lengths inside the parentheses and multiply by their lengths outside, which gives us

$$\left(\frac{1}{\sqrt{3}}|a_0\rangle + \frac{\sqrt{2}}{\sqrt{3}}|a_1\rangle\right)\frac{\sqrt{3}}{2}|b_0\rangle + (1|a_0\rangle + 0|a_1\rangle)\frac{1}{2}|b_1\rangle.$$

When Bob measures his qubit he gets 0 with probability 3/4 and 1 with probability 1/4. When Bob gets 0, Alice's qubit jumps to state $\left(\frac{1}{\sqrt{3}}|a_0\rangle + \frac{\sqrt{2}}{\sqrt{3}}|a_1\rangle\right)$. When he gets 1, her qubit becomes $|a_0\rangle$.

When the first person measures her or his qubit, the second person's qubit immediately jumps to one of two states. These states depend on the result of the first person's measurement. This is quite unlike our everyday experience. Later we will see clever ways of exploiting entangled qubits, but first we consider superluminal communication.

Superluminal Communication

Superluminal communication is communication faster than the speed of light. Two apparently contradictory inferences seem to be able to be deduced concerning this. The first is that Einstein's special theory of relativity tells us that as you travel faster, approaching the speed of light, time slows down. If you could travel at the speed of light, time stops. And if you travel faster than the speed of light, time would go backward. The theory also tells us that as we approach the speed of light our mass increases without bound, which means that we can never reach that speed. Also, it seems unlikely that we could go back in time. If we could, then we run into all the science fiction scenarios in which we can prevent some history-changing event from happening. Time travel seems to lead to contradictions. It's not just physical travel that seems to be ruled out, but also communication. If we could send messages back in time, we could still change the course of history—we could still design scenarios in which the communication causes some dramatic change in the present—for example, prevents us being born. So, one of our immediate thoughts is that superluminal communication should not be possible.

On the other hand, suppose that Alice and Bob are on opposite sides of the universe and have a number of entangled qubits. These are electrons whose spin states are entangled. Alice has one of each entangled pair of electrons in her possession, and Bob has the other one. (Though we are talking about entangled electrons, we should be clear that the actual electrons are totally separate. It's their spin states that are entangled.)

When Alice makes a measurement on one of her electrons, the spin state of the corresponding electron in Bob's possession instantaneously jumps into one of two distinct states. Instantaneous is clearly faster than the speed of light! Can't entanglement be used for instantaneous communication?

Let's suppose that each pair of the entangled electrons is in the entangled spin state that we have just studied:

$$\frac{1}{2}|a_0\rangle|b_0\rangle + \frac{1}{2}|a_0\rangle|b_1\rangle + \frac{1}{\sqrt{2}}|a_1\rangle|b_0\rangle + 0|a_1\rangle|b_1\rangle.$$

Suppose that Alice measures the spins of her electrons before Bob measures the spin of their partners. We have seen that she gets a random string of 0s and 1s, with each symbol occurring with equal probability.

Suppose instead that Bob measures his spins before Alice. Then Alice measures the spins. What does she now get? When Alice makes her measurement, they will both have made measurements, so we can use the probability amplitudes of the initial expression. We know that they will obtain 00 and 01 with probability 1/4, 10 with probability 1/2 and 11 with probability 0. Consequently, Alice will get 0 with probability of $\frac{1}{4} + \frac{1}{4} = \frac{1}{2}$, and 1 with probability $\frac{1}{2} + 0 = \frac{1}{2}$. So, Alice gets a random string of 0s and 1s, with each symbol occurring with equal probability. But this is exactly the same situation as when she measured first. So Alice cannot tell from her measurements whether they were made before or after Bob's. All entangled states behave this way. If there is no way of Alice and Bob being able to tell from their measurements who went first, there certainly can be no way of sending any information from one to the other.

We have shown that Alice and Bob cannot send information when their qubits have a particular entangled state, but the argument generalizes to any entangled state. No matter what states Alice's and Bob's qubits have, it is impossible for them to send information by solely measuring their qubits.

Now that we have seen that superluminal communication is not possible, we turn to the more prosaic task of writing tensor products using standard bases. But afterward we will return to our exploration of entangled qubits using the quantum clock example from the previous chapters.

The Standard Basis for Tensor Products

The standard basis for $\mathbb{R}^2$ is $\left(\begin{bmatrix} 1 \\ 0 \end{bmatrix}, \begin{bmatrix} 0 \\ 1 \end{bmatrix} \right)$. When both Alice and Bob use the standard basis, tensor products have the form:

$$r\begin{bmatrix} 1 \\ 0 \end{bmatrix} \otimes \begin{bmatrix} 1 \\ 0 \end{bmatrix} + s\begin{bmatrix} 1 \\ 0 \end{bmatrix} \otimes \begin{bmatrix} 0 \\ 1 \end{bmatrix} + t\begin{bmatrix} 0 \\ 1 \end{bmatrix} \otimes \begin{bmatrix} 1 \\ 0 \end{bmatrix} + u\begin{bmatrix} 0 \\ 1 \end{bmatrix} \otimes \begin{bmatrix} 0 \\ 1 \end{bmatrix}$$

Therefore, the standard ordered basis for $\mathbb{R}^2 \otimes \mathbb{R}^2$ is

$$\left(\begin{bmatrix} 1 \\ 0 \end{bmatrix} \otimes \begin{bmatrix} 1 \\ 0 \end{bmatrix}, \begin{bmatrix} 1 \\ 0 \end{bmatrix} \otimes \begin{bmatrix} 0 \\ 1 \end{bmatrix}, \begin{bmatrix} 0 \\ 1 \end{bmatrix} \otimes \begin{bmatrix} 1 \\ 0 \end{bmatrix}, \begin{bmatrix} 0 \\ 1 \end{bmatrix} \otimes \begin{bmatrix} 0 \\ 1 \end{bmatrix} \right).$$

Since it has four vectors in the basis, it is a four-dimensional space. The standard four-dimensional space is $\mathbb{R}^4$ with ordered basis is

$$\left(\begin{bmatrix} 1 \\ 0 \\ 0 \\ 0 \end{bmatrix}, \begin{bmatrix} 0 \\ 1 \\ 0 \\ 0 \end{bmatrix}, \begin{bmatrix} 0 \\ 0 \\ 1 \\ 0 \end{bmatrix}, \begin{bmatrix} 0 \\ 0 \\ 0 \\ 1 \end{bmatrix} \right).$$

We identify the basis vectors in $\mathbb{R}^2 \otimes \mathbb{R}^2$ with those in $\mathbb{R}^4$, making sure to respect the ordering.

$$\begin{bmatrix} 1 \\ 0 \\ 0 \\ 0 \end{bmatrix} = \begin{bmatrix} 1 \\ 0 \end{bmatrix} \otimes \begin{bmatrix} 1 \\ 0 \end{bmatrix}, \quad \begin{bmatrix} 0 \\ 1 \\ 0 \\ 0 \end{bmatrix} = \begin{bmatrix} 1 \\ 0 \end{bmatrix} \otimes \begin{bmatrix} 0 \\ 1 \end{bmatrix}, \quad \begin{bmatrix} 0 \\ 0 \\ 1 \\ 0 \end{bmatrix} = \begin{bmatrix} 0 \\ 1 \end{bmatrix} \otimes \begin{bmatrix} 1 \\ 0 \end{bmatrix}, \quad \begin{bmatrix} 0 \\ 0 \\ 0 \\ 1 \end{bmatrix} = \begin{bmatrix} 0 \\ 1 \end{bmatrix} \otimes \begin{bmatrix} 0 \\ 1 \end{bmatrix}.$$

The easiest way to remember this is by the following construction.

$$\begin{bmatrix} a_0 \\ a_1 \end{bmatrix} \otimes \begin{bmatrix} b_0 \\ b_1 \end{bmatrix} = \begin{bmatrix} a_0 \begin{bmatrix} b_0 \\ b_1 \end{bmatrix} \\ a_1 \begin{bmatrix} b_0 \\ b_1 \end{bmatrix} \end{bmatrix} = \begin{bmatrix} a_0 b_0 \\ a_0 b_1 \\ a_1 b_0 \\ a_1 b_1 \end{bmatrix}.$$

Notice also that the subscripts follow the standard binary ordering: 00, 01, 10, 11.

How Do You Entangle Qubits?

This book is about the mathematics that underlies quantum computing. It is not about how to physically create a quantum computer. We are not going to spend much time on the details of physical experiments, but the question of how physicists create entangled particles is such an important one that we will briefly address it. We can represent entangled qubits by either entangled photons or electrons. Though we often say the particles are entangled, what we really mean is that the vector describing their states, a tensor in $\mathbb{R}^2 \otimes \mathbb{R}^2$, is entangled. The actual particles are separate and, as we have just noted, can be very far apart. That said, the question remains: How do you go about creating a pair of particles whose state vector is entangled? First, we look at how physical experiments create entangled particles. Then we look at how quantum gates create entangled qubits.

The most commonly used method at this time involves photons. The process is called *spontaneous parametric down-conversion*. A laser beam sends photons through a special crystal. Most of the photons just pass through, but some photons split into two. Energy and momentum must be conserved—the total energy and momentum of the two resulting photons must equal the energy and momentum of the initial photon. The conservation laws guarantee that the state describing the polarization of the two photons is entangled.

In the universe, electrons are often entangled. At the start of the book we described Stern and Gerlach's experiment on silver atoms. Recall that the electron spins in the inner orbits canceled, leaving the lone electron in the outer orbit to give its spin to the atom. The innermost orbit has two electrons. These are entangled so that their spins cancel. We can think of the state vector describing the spin of these electrons as

$$\frac{1}{\sqrt{2}}\begin{bmatrix}1\\0\end{bmatrix}\otimes\begin{bmatrix}0\\1\end{bmatrix}-\frac{1}{\sqrt{2}}\begin{bmatrix}0\\1\end{bmatrix}\otimes\begin{bmatrix}1\\0\end{bmatrix}.$$

Entangled electrons also occur in superconductors, and these electrons have been used in experiments. However, often we want to have entangled particles that are far apart—as we will see later when we talk about the Bell test.

The main problem with using entangled electrons that are near one another and then separating them is that they have a tendency to interact with the environment. It is difficult to separate them without this happening. On the other hand, entangled photons are much easier to separate, though more difficult to measure. It is possible, however, to get the best of both worlds. This has been done by an international team based at the Delft University of Technology in what they describe as a loophole free Bell test. They used two diamonds separated by 1.3 kilometers. Each diamond had slight imperfections—nitrogen atoms altered the carbon atom lattice structure in places. Electrons become trapped in the defects. A laser excited an electron in each of the diamonds in such a way that both electrons emitted photons. The emitted photons were entangled with the spins of the electrons that they were emitted from. The photons then traveled toward one another through a fiber optic cable and met in a beam splitter—a standard piece of equipment that is usually used to split a beam of photons in two, but here it is used to entangle the two photons. The photons were then measured. The result was that the two electrons were now entangled with one another.* (We will explain why the team was doing this experiment in the next chapter.)

In quantum computing, we will usually input unentangled qubits and entangle them using the *CNOT* gate. Later we will explain exactly what gates are, but the actual computations involve just matrix multiplication. We briefly look at this.

* There is a short video on this at https://www.youtube.com/watch?v=AE8Ma QJkRcg/.

Using the *CNOT* Gate to Entangle Qubits

We leave the actual definition of what a quantum gate is until later, but we'll just make the comment now that they correspond to orthonormal bases or, equivalently, to orthogonal matrices.

The standard ordered basis for four-dimensional space is $\mathbb{R}^4$ is

$$\left(\begin{bmatrix} 1 \\ 0 \\ 0 \\ 0 \end{bmatrix}, \begin{bmatrix} 0 \\ 1 \\ 0 \\ 0 \end{bmatrix}, \begin{bmatrix} 0 \\ 0 \\ 1 \\ 0 \end{bmatrix}, \begin{bmatrix} 0 \\ 0 \\ 0 \\ 1 \end{bmatrix} \right).$$

The *CNOT* gate comes from interchanging the order of the last two elements. This results in the matrix for the *CNOT* gate.

$$\begin{bmatrix} 1 & 0 & 0 & 0 \\ 0 & 1 & 0 & 0 \\ 0 & 0 & 0 & 1 \\ 0 & 0 & 1 & 0 \end{bmatrix}$$

This gate acts on pairs of qubits. To use the matrix, everything must be written using four-dimensional vectors. We look at an example.

We start by taking the unentangled tensor product

$$\frac{1}{\sqrt{2}} \begin{bmatrix} 1 \\ 1 \end{bmatrix} \otimes \begin{bmatrix} 1 \\ 0 \end{bmatrix} = \frac{1}{\sqrt{2}} \begin{bmatrix} 1 \\ 0 \\ 1 \\ 0 \end{bmatrix}.$$

When we send qubits through the gate, they are changed. The resulting qubits are obtained by multiplying by the matrix.

$$\begin{bmatrix} 1 & 0 & 0 & 0 \\ 0 & 1 & 0 & 0 \\ 0 & 0 & 0 & 1 \\ 0 & 0 & 1 & 0 \end{bmatrix} \begin{bmatrix} \frac{1}{\sqrt{2}} \\ 0 \\ \frac{1}{\sqrt{2}} \\ 0 \end{bmatrix} = \begin{bmatrix} \frac{1}{\sqrt{2}} \\ 0 \\ 0 \\ \frac{1}{\sqrt{2}} \end{bmatrix} = \frac{1}{\sqrt{2}} \begin{bmatrix} 1 \\ 0 \\ 0 \\ 1 \end{bmatrix}.$$

This last vector corresponds to a pair of entangled qubits—the product of the inner amplitudes is zero, which is not equal to the product of the outer amplitudes. This can be rewritten as

$$\frac{1}{\sqrt{2}}\begin{bmatrix} 1 \\ 0 \end{bmatrix} \otimes \begin{bmatrix} 1 \\ 0 \end{bmatrix} + \frac{1}{\sqrt{2}}\begin{bmatrix} 0 \\ 1 \end{bmatrix} \otimes \begin{bmatrix} 0 \\ 1 \end{bmatrix}.$$

We will often use entangled qubits in this state. It has the very nice property that if Alice and Bob measure in the standard basis, they will both get $\begin{bmatrix} 1 \\ 0 \end{bmatrix}$, corresponding to 0, or they will both get $\begin{bmatrix} 0 \\ 1 \end{bmatrix}$, corresponding to 1. The two cases are equally probable.**

We examine this further with a quantum clock analogy.

Entangled Quantum Clocks

Recall the quantum clock metaphor. We can ask only about whether the hand is pointing in a certain direction, and the clock will answer either that it is or that it is pointing in the opposite direction.

We let the vector $\begin{bmatrix} 1 \\ 0 \end{bmatrix}$ correspond to pointing to twelve, and $\begin{bmatrix} 0 \\ 1 \end{bmatrix}$ to pointing to six. Consider a pair of clocks in the entangled state $\frac{1}{\sqrt{2}}\begin{bmatrix} 1 \\ 0 \end{bmatrix} \otimes \begin{bmatrix} 1 \\ 0 \end{bmatrix} + \frac{1}{\sqrt{2}}\begin{bmatrix} 0 \\ 1 \end{bmatrix} \otimes \begin{bmatrix} 0 \\ 1 \end{bmatrix}$. In fact, consider one hundred pairs of clocks, each pair of which is in this state. Suppose that you have one hundred of these clocks, and I have the hundred partners. We are both going to ask the same question repeatedly: Is the hand pointing toward twelve?

In the first scenario, we don't contact one another. We just go through the clocks one at a time and ask the question. Each time the clock will answer either yes or no. We will write 1 if it is yes, and 0 if it is no. After we have finished asking questions, we have a string of 0s and 1s. I analyze my string and you analyze yours. Both strings are a random sequence of 0s and 1s. Both digits occur about the same number of times. We now contact one another and compare strings. Both your string and my string are identical. In all one hundred places the strings agree.

In the second scenario, we again each have one hundred clocks. This time we make an agreement that you will measure first. You will ask your question on the hour, and I will ask mine half an hour later. During these half-hours between our questions you will call me and tell me what my clock's

***

** In the next chapter, we will see that Alice and Bob don't need to stick to the standard basis. If they both use the same orthonormal basis, no matter which one, they will still get exactly the same results.

answer will be. At the end of the experiment we both have a string of 0s and 1s. Both strings agree in every place. Every time you called me and told me what the result of my measurement was going to be you were exactly right. Can we conclude that your measurements were affecting mine?

Well, suppose that I now tell you that I was cheating. I didn't follow the rules. In fact, I was asking the questions of the clock half an hour before you asked yours. I knew your answer before you did. Your calls were just confirming what I knew.

There is no way from the data that you can tell whether or not I was following the rules or I was cheating. There is no way you can tell whether I am asking my questions before or after you asked yours.

There is no causation here, just correlation. As we saw earlier, we cannot use these entangled clocks to send messages between us. But the process is still mysterious. Albert Einstein described entanglement as implying spooky action at a distance. Nowadays many people would say that there is no action, just correlation. Of course, we can quibble about the definition of "action," but even if we agree that there is no action, there seems to be something spooky going on.

Suppose that you and I have a pair of the entangled quantum clocks, and we are talking on the phone to one another. Neither of us has asked our clock a question, so they are still entangled. In this state, if you were to ask your clock the question, you would have an equal chance of getting an answer that the hand was pointing to twelve or six. But as soon as I ask my clock a question, you no longer have an equal chance of getting one of the two answers. You will get exactly the same answer as mine.

This correlation would not be spooky if when our clocks were entangled it was decided, but unknown to us, whether both our hands were pointing at either twelve or six. We had to wait until one of us asked the question, and as soon as one of us knows the answer so does the other.

But this is not what our model describes. Our model says that the decision on which direction our hands are pointing is not made beforehand. It's made only when the first of us asks our question. This is what makes it spooky.

In the next chapter we will look at this in detail. We will look at a model that incorporates correlation in an intuitive and nonspooky way. Unfortunately, it is wrong. John Stewart Bell came up with an ingenious test that shows that the simple explanation is not correct and that the mysterious spookiness has to remain.

5 Bell's Inequality

We have seen a mathematical model of a small portion of quantum mechanics that concerns the spin of particles or the polarization of photons, and that gives us the mathematics describing qubits. This is the standard model, often called the *Copenhagen interpretation* after the city where Niels Bohr was living and working.

Some of the great physicists of the early twentieth century, including Albert Einstein and Erwin Schrödinger, didn't like this model, with its interpretation of states jumping with given probabilities to basis states. They objected to both the use of probability and to the concept of action at a distance. They thought that there should be a better model using "hidden variables" and "local realism." They weren't objecting to using the Copenhagen model for doing calculations, but they thought that there should be a deeper theory that would explain why the calculations were producing correct answers—a theory that eliminated the randomness and explained the mystery.

Bohr and Einstein were both interested in the philosophy of quantum mechanics and had a series of debates about the true meaning of the theory. In this chapter we will look at their two different viewpoints. You might be wondering if we are digressing and that the philosophical underpinnings are not necessary to understand quantum computation. We all now know that Einstein and Schrödinger's view was wrong and that the Copenhagen model is regarded as the standard description. But Einstein and Schrödinger were both brilliant scientists, and there are a number of reasons to study their arguments.

The first reason is that the debates between Bohr and Einstein focused on local realism. We will explain more about this in a moment, but essentially local realism means that a particle can only be influenced by something changing in its vicinity. Practically all of us are local realists, but quantum mechanics shows us that we are wrong. Einstein's model seems to us to be the natural and correct model—at least it does to me. When I first heard of quantum entanglement, my natural assumption was to assume a model similar to Einstein's. You too might be thinking about entanglement incorrectly. These arguments are important to the philosophy of physics and help us to understand that the mysteriousness cannot be eliminated.

John Stewart Bell was an Irish physicist. He devised an ingenious test that could distinguish between the two models. Many were surprised that the models were not just philosophies, but testable theories. We have learned only a small portion of the mathematics needed for quantum mechanics, but it is exactly what is needed to understand Bell's result. His test has been carried out several times. It is tricky to eliminate all possible biases in the setup of this experiment, but more and more possible loopholes have been excluded. The results have always been in accordance with the Copenhagen interpretation. Since Bell's result is one of the most important of the twentieth century and we have the mathematical machinery lined up, it makes sense to look at it.

You may still be wondering what any of this has to do with quantum computation. We will see at the end of this chapter that the ideas behind Bell's inequality can be used for sending encrypted messages. Also, the entangled qubits that Bell uses will reappear when we look at quantum algorithms. So this chapter has connections to quantum computation. But the main reason for this chapter is that I find this material fascinating, and I hope you will too.

We start by looking at the entangled qubits we introduced in the last chapter and see what happens if we measure them in different bases. We begin our analysis using the standard model—the Copenhagen model—that we have seen in the earlier chapters.

Entangled Qubits in Different Bases

In the last chapter we looked at two entangled clocks in the state

$$\frac{1}{\sqrt{2}}\begin{bmatrix}1\\0\end{bmatrix}\otimes\begin{bmatrix}1\\0\end{bmatrix}+\frac{1}{\sqrt{2}}\begin{bmatrix}0\\1\end{bmatrix}\otimes\begin{bmatrix}0\\1\end{bmatrix}.$$

We observed that if Alice and Bob each had one of the clocks, and both asked whether the hand was pointing toward twelve, both would either get the answer that it was or that it was pointing toward six. Both possibilities were equally likely, but both Alice and Bob get exactly the same answer. We now ask what happens if Alice and Bob change the direction in which they are measuring. For example, what happens if they both ask whether the hands are pointing to four? We know that the clocks will answer that the hands are pointing either toward four or to ten, but will Alice and Bob get exactly the same answer? Are both answers equally likely?

First, we give an intuitive argument for two qubits in the entangled state

$$\frac{1}{\sqrt{2}}\begin{bmatrix}1\\0\end{bmatrix}\otimes\begin{bmatrix}0\\1\end{bmatrix}+\frac{1}{\sqrt{2}}\begin{bmatrix}0\\1\end{bmatrix}\otimes\begin{bmatrix}1\\0\end{bmatrix}.$$

Two electrons might represent this state. Suppose that Alice and Bob measure the spin of their electrons in direction $0°$. If Alice gets N, Bob gets S. If Alice gets S, Bob gets N. As we mentioned earlier, this might represent two electrons in an atom where the spins cancel. But we would expect the spins to cancel in every direction, so we would expect that if Alice and Bob chose a new basis for measurements they would still get spins in the opposite direction. Symmetry also seems to imply that both directions should be equally likely.

This intuitive argument leads us to conjecture that if we have entangled qubits in the state

$$\frac{1}{\sqrt{2}}\begin{bmatrix}1\\0\end{bmatrix}\otimes\begin{bmatrix}1\\0\end{bmatrix}+\frac{1}{\sqrt{2}}\begin{bmatrix}0\\1\end{bmatrix}\otimes\begin{bmatrix}0\\1\end{bmatrix}$$

and then rewrite this state using a new orthonormal basis $(|b_0\rangle,|b_1\rangle)$, we ought to get $\frac{1}{\sqrt{2}}|b_0\rangle\otimes|b_0\rangle+\frac{1}{\sqrt{2}}|b_1\rangle\otimes|b_1\rangle$. Of course, our argument is intuitive and clearly making intuitive arguments about something as counterintuitive as quantum mechanics is not entirely persuasive, but in this case we are correct, as we will now prove.

Proof That $\frac{1}{\sqrt{2}}\begin{bmatrix}1\\0\end{bmatrix}\otimes\begin{bmatrix}1\\0\end{bmatrix}+\frac{1}{\sqrt{2}}\begin{bmatrix}0\\1\end{bmatrix}\otimes\begin{bmatrix}0\\1\end{bmatrix}$ **Equals** $\frac{1}{\sqrt{2}}|b_0\rangle\otimes|b_0\rangle+\frac{1}{\sqrt{2}}|b_1\rangle\otimes|b_1\rangle$

We start by writing the kets $|b_0\rangle$ and $|b_1\rangle$ as column vectors. We let $|b_0\rangle=\begin{bmatrix}a\\b\end{bmatrix}$ and $|b_1\rangle=\begin{bmatrix}c\\d\end{bmatrix}$. Next we express our standard basis vectors as linear combinations of the new basis vectors. We do this in the standard way (using the second tool at the end of chapter 2). We start with $\begin{bmatrix} 1 \\ 0 \end{bmatrix}$. The equation

$$\begin{bmatrix} a & b \\ c & d \end{bmatrix}\begin{bmatrix} 1 \\ 0 \end{bmatrix} = \begin{bmatrix} a \\ c \end{bmatrix}$$

tells us that

$$\begin{bmatrix} 1 \\ 0 \end{bmatrix} = a\begin{bmatrix} a \\ b \end{bmatrix} + c\begin{bmatrix} c \\ d \end{bmatrix}.$$

Consequently,

$$\begin{bmatrix} 1 \\ 0 \end{bmatrix} \otimes \begin{bmatrix} 1 \\ 0 \end{bmatrix} = \left(a\begin{bmatrix} a \\ b \end{bmatrix} + c\begin{bmatrix} c \\ d \end{bmatrix} \right) \otimes \begin{bmatrix} 1 \\ 0 \end{bmatrix}.$$

Rearranging the terms on the right gives

$$a\begin{bmatrix} a \\ b \end{bmatrix} \otimes \begin{bmatrix} 1 \\ 0 \end{bmatrix} + c\begin{bmatrix} c \\ d \end{bmatrix} \otimes \begin{bmatrix} 1 \\ 0 \end{bmatrix},$$

which can be rewritten as

$$\begin{bmatrix} a \\ b \end{bmatrix} \otimes \begin{bmatrix} a \\ 0 \end{bmatrix} + \begin{bmatrix} c \\ d \end{bmatrix} \otimes \begin{bmatrix} c \\ 0 \end{bmatrix}.$$

Thus, $\begin{bmatrix} 1 \\ 0 \end{bmatrix} \otimes \begin{bmatrix} 1 \\ 0 \end{bmatrix} = \begin{bmatrix} a \\ b \end{bmatrix} \otimes \begin{bmatrix} a \\ 0 \end{bmatrix} + \begin{bmatrix} c \\ d \end{bmatrix} \otimes \begin{bmatrix} c \\ 0 \end{bmatrix}.$

A similar calculation shows that

$$\begin{bmatrix} 0 \\ 1 \end{bmatrix} \otimes \begin{bmatrix} 0 \\ 1 \end{bmatrix} = \begin{bmatrix} a \\ b \end{bmatrix} \otimes \begin{bmatrix} 0 \\ b \end{bmatrix} + \begin{bmatrix} c \\ d \end{bmatrix} \otimes \begin{bmatrix} 0 \\ d \end{bmatrix}.$$

Adding these two results gives us

$$\begin{bmatrix} 1 \\ 0 \end{bmatrix} \otimes \begin{bmatrix} 1 \\ 0 \end{bmatrix} + \begin{bmatrix} 0 \\ 1 \end{bmatrix} \otimes \begin{bmatrix} 0 \\ 1 \end{bmatrix} = \begin{bmatrix} a \\ b \end{bmatrix} \otimes \left(\begin{bmatrix} a \\ 0 \end{bmatrix} + \begin{bmatrix} 0 \\ b \end{bmatrix} \right) + \begin{bmatrix} c \\ d \end{bmatrix} \otimes \left(\begin{bmatrix} c \\ 0 \end{bmatrix} + \begin{bmatrix} 0 \\ d \end{bmatrix} \right).$$

This simplifies to

$$\begin{bmatrix} a \\ b \end{bmatrix} \otimes \begin{bmatrix} a \\ b \end{bmatrix} + \begin{bmatrix} c \\ d \end{bmatrix} \otimes \begin{bmatrix} c \\ d \end{bmatrix},$$

which is just $|b_0\rangle \otimes |b_0\rangle + |b_1\rangle \otimes |b_1\rangle$.

So $\frac{1}{\sqrt{2}}\begin{bmatrix}1\\0\end{bmatrix}\otimes\begin{bmatrix}1\\0\end{bmatrix}+\frac{1}{\sqrt{2}}\begin{bmatrix}0\\1\end{bmatrix}\otimes\begin{bmatrix}0\\1\end{bmatrix}$ does equal $\frac{1}{\sqrt{2}}|b_0\rangle\otimes|b_0\rangle+\frac{1}{\sqrt{2}}|b_1\rangle\otimes|b_1\rangle$.

This result tells us that if Alice and Bob have qubits that are entangled with state $\frac{1}{\sqrt{2}}\begin{bmatrix}1\\0\end{bmatrix}\otimes\begin{bmatrix}1\\0\end{bmatrix}+\frac{1}{\sqrt{2}}\begin{bmatrix}0\\1\end{bmatrix}\otimes\begin{bmatrix}0\\1\end{bmatrix}$, and if they both choose to measure their qubits with respect to an orthonormal basis $(|b_0\rangle,|b_1\rangle)$, the entangled state can be rewritten as $\frac{1}{\sqrt{2}}|b_0\rangle|b_0\rangle+\frac{1}{\sqrt{2}}|b_1\rangle|b_1\rangle$. When the first measurement is made, the state jumps to either $|b_0\rangle|b_0\rangle$ or to $|b_1\rangle|b_1\rangle$, where both of these now unentangled states are equally likely to occur. The consequence is that when Alice and Bob have both measured their qubits they will both get 0 or they will both get 1, and both outcomes are equally likely.

For Bell's result we want to measure the entangled qubits using three different bases. These are the bases that correspond to rotating our measuring device through 0°, 120°, and 240°. For our entangled clocks, we are asking one of three questions, whether the hand is pointing to twelve, to four, or to eight. If we denote these bases by $(|\uparrow\rangle,|\downarrow\rangle)$, $(|\searrow\rangle,|\nwarrow\rangle)$, and $(|\swarrow\rangle,|\nearrow\rangle)$, then the following are three descriptions of exactly the same entangled state:

$$\frac{1}{\sqrt{2}}|\uparrow\rangle|\uparrow\rangle+\frac{1}{\sqrt{2}}|\downarrow\rangle|\downarrow\rangle \qquad \frac{1}{\sqrt{2}}|\searrow\rangle|\searrow\rangle+\frac{1}{\sqrt{2}}|\nwarrow\rangle|\nwarrow\rangle \qquad \frac{1}{\sqrt{2}}|\swarrow\rangle|\swarrow\rangle+\frac{1}{\sqrt{2}}|\nearrow\rangle|\nearrow\rangle$$

We now turn to Einstein and see how he viewed these entangled states.

Einstein and Local Realism

Gravity provides a good example to explain local realism. Newton's law of gravity gives a formula that tells us the strength of the force between two masses. If you plug in the size of the masses, the distance they are apart, and the gravitational constant, the formula gives the size of the attractive force. Newton's law transformed physics. It can be used, for example, to show that a planet orbiting a star moves in an elliptical orbit. But though it tells us the value of the force, it does not tell us the mechanism that connects the planet to the sun.

Although Newton's law of gravitation was useful for calculations, it did not explain how gravity worked. Newton, himself, was concerned about this. Everyone thought that there should be some deeper theory that explained the action of gravity. Various proposals were made, often involving an "aether" that was supposed to permeate the universe. Though there was no consensus on how the mechanism behind gravity worked, there was consensus that gravity was not spooky action at a distance and that some explanation would be found. There was a belief in what we now call local realism.

Newton's law of gravitation was superseded by Einstein's general theory of gravitation. Einstein's theory not only improved on Newton's in terms of accurately predicting astronomical observations that could not be deduced using Newton's theory, but it also gave an explanation as to how gravity worked. It described the warping of space-time. A planet moved according to the shape of space-time where it was located. There was no spooky action at a distance. Einstein's theory was not only more precise, but it also gave a description of how gravity worked, and this description was local. A planet moves according to the shape of space in its vicinity.

The Copenhagen interpretation of quantum mechanics, of course, reintroduced this idea of spooky action at a distance. When you measure a pair of entangled qubits the state immediately changes, even if the qubits are physically far apart. Einstein's philosophy seems entirely natural. He had just eliminated spooky action from the theory of gravity, and now it was being proposed again. The difference now was that Bohr didn't believe that there was some deeper theory that could explain the mechanism behind this action. Einstein disagreed.

Einstein believed he could prove that Bohr was wrong. With Boris Podolsky and Nathan Rosen, he wrote a paper pointing out that his special theory of relativity implied that information could not travel faster than the speed of light, but instantaneous action at a distance would mean that information could be sent from Alice to Bob instantaneously. This problem became known as the *EPR paradox*, for Einstein-Podolsky-Rosen.

Nowadays, the EPR paradox is usually described in terms of spin, and this is how we will do it, but this was not how Einstein et al. described the problem. They considered position and momentum of two entangled particles. It was David Bohm who reformulated the problem in terms of spin. Bohm's formulation is the one that is practically always used now, and it is the formulation that John Stewart Bell used to calculate his important inequality. Even though Bohm played a major role in describing and reformulating the paradox, his name is usually omitted.

In the last chapter we pointed out that the Copenhagen interpretation does not allow information to be transmitted faster than the speed of light, and so although the EPR paradox is not really a paradox, there is still the question of whether there can be an explanation that eliminates the spooky action.

Einstein and Hidden Variables

In the classical view, physics is deterministic—if you know all the initial conditions to infinite precision, then you can predict the outcome with certainty. Of course, you can only know initial conditions to some finite precision, meaning that there will always be some small error in what is measured—a small difference between the measured value and the true value. As time progresses this error can grow until we are unable to make any sensible prediction for what happens in the long run. This idea forms the basis for what is commonly known as sensitive dependence on initial conditions. It explains why forecasting the weather for more than a week or so is very unreliable. It is important to remember, however, that the underlying theory is deterministic. The weather seems unpredictable, but this is not due to some inherent randomness, it is just that we cannot make measurements that are sufficiently accurate.

Another area where probability is incorporated into classical physics is in laws concerning gases—the laws of thermodynamics—but again the underlying theory is still deterministic. If we know exactly the velocities and masses of every molecule in the gas, in theory we can predict with complete accuracy what happens to each molecule in the future. In practice, of course, there are far too many molecules to consider them one by one, and so we take average values and look at the gas from a statistical viewpoint.

This classical, deterministic view was what Einstein was referring to when he famously said that God does not play dice with the universe. He felt that the use of probability in quantum mechanics showed that the theory was not complete. There should be a deeper theory, perhaps involving new variables, that is deterministic but looks probabilistic if you don't consider all of these as yet unknown variables. These as yet unknown variables became known as hidden variables.

A Classical Explanation of Entanglement

We begin with our quantum clocks in state $\frac{1}{\sqrt{2}}|\uparrow\rangle|\uparrow\rangle + \frac{1}{\sqrt{2}}|\downarrow\rangle|\downarrow\rangle$. Alice and Bob are going to ask the question about whether the hand is pointing to twelve. The quantum model says that Alice and Bob will get exactly the same answer: that it's pointing to twelve or it's pointing to six. Both answers are equally likely. We can actually perform experiments measuring the spin of entangled electrons. The experimental outcomes are exactly what the quantum model predicts. How do we explain this with a classical model?

The classical interpretation for the preceding situation is quite simple. Electrons have a definite spin in any direction. Entangled electrons become entangled through some local interaction. Again, we appeal to hidden variables and a deeper theory. We don't know exactly what happens, but there is some local process that puts the electrons in exactly the same spin configuration state. When they are entangled, a direction of spin is chosen for both electrons.

This can be compared to our being given a deck of cards that we shuffle. We then take out one card without looking at it. We cut the card in two and put the halves in two envelopes, all the time without any knowledge of which card has been chosen. We then send the cards to Bob and Alice, who live in different parts of the universe. Alice and Bob have no idea which card they have. It could be anyone of the fifty-two, but as soon as Alice opens her envelope and sees the jack of diamonds she knows that Bob's card is also the jack of diamonds. There is no action at a distance, and there is nothing spooky going on.

For Bell's result, we need to measure our entangled qubits in three different directions. We return to our entangled clock analogy. We will be asking one of three questions, about whether the hand is pointing to twelve, to four, or to eight. The quantum theoretical model predicts that for each question the answer will be either that the hand is pointing in the direction asked or that it is pointing in the opposite direction. For each question both answers are equally likely. But when Alice and Bob ask exactly the same question, they will both get exactly the same answer. We can describe this classically by giving essentially the same answer as before.

There is some local process that entangles the clocks. We don't attempt to describe exactly how this is done, but just appeal to hidden variables—there is some deeper theory that explains it. But when the clocks are entangled, definite answers to the three questions are chosen. This can be compared to our having three decks of cards, each with different colored backs. We take a card from the blue deck, from the red deck and the green deck. We cut these three cards in half and mail three halves to Alice and the other three halves to Bob. If Alice looks at her green card and sees the jack of diamonds, she knows that Bob's green card is also the jack of diamonds.

For our quantum clocks, the classical theory says that there is a definite answer to each question that is already determined before we ask it. Quantum theory says, contrarily, that the answer to the question is not determined up until the time we ask it.

Bell's Inequality

Imagine that we are generating a stream of pairs of qubits that we are sending to Alice and Bob. Each pair of qubits is in the entangled state $\frac{1}{\sqrt{2}}|\uparrow\rangle|\uparrow\rangle + \frac{1}{\sqrt{2}}|\downarrow\rangle|\downarrow\rangle$. Alice randomly chooses to measure her qubit in direction 0°, 120°, or 240°. Each of these directions is chosen randomly, each with probability 1/3. Alice doesn't bother to keep track of the directions she has chosen, but she does write down whether she gets 0 or 1 as the answer. (Remember, 0 corresponds to the first basis vector and 1 to the second.) Shortly after Alice has measured her qubit, Bob chooses one of the same three directions at random, each with probability 1/3, and measures his qubit. Like Alice, he doesn't record the direction, just the result of whether he obtained either 0 or 1.

In this way, both Alice and Bob generate a long string of 0s and 1s. They then compare their strings symbol by symbol. If they agree on the first symbol, they write down A. If they disagree on the first symbol, they write down D. Then they look at the second symbol and write down A or D depending on whether the symbols agree or disagree. They continue in this way through their entire strings.

In this way they generate a new string consisting of As and Ds. What proportion of the string is made up of As? Bell realized that the quantum mechanics model and the classical model gave different numbers for the answer.

The Answer of Quantum Mechanics

The qubits are in the entangled spin state $\frac{1}{\sqrt{2}}|\uparrow\rangle|\uparrow\rangle + \frac{1}{\sqrt{2}}|\downarrow\rangle|\downarrow\rangle$. We have already observed that if Alice and Bob both choose the same measurement direction, then they will get the same answer. The question is what happens if they choose different bases.

We will take the case when Alice chooses $(|\searrow\rangle, |\nwarrow\rangle)$ and Bob chooses $(|\swarrow\rangle, |\nearrow\rangle)$. The entangled state is $\frac{1}{\sqrt{2}}|\uparrow\rangle|\uparrow\rangle + \frac{1}{\sqrt{2}}|\downarrow\rangle|\downarrow\rangle$, which can be written in Alice's basis as $\frac{1}{\sqrt{2}}|\searrow\rangle|\searrow\rangle + \frac{1}{\sqrt{2}}|\nwarrow\rangle|\nwarrow\rangle$. When Alice makes her measurement, the state jumps to either $|\searrow\rangle|\searrow\rangle$ or $|\nwarrow\rangle|\nwarrow\rangle$; both are equally likely. If it jumps to $|\searrow\rangle|\searrow\rangle$, she will write down 0. If it jumps to $|\nwarrow\rangle|\nwarrow\rangle$, she will write down 1.

Bob must now make his measurement. Suppose after Alice's measurement that the qubits are in state $|\searrow\rangle|\searrow\rangle$, so Bob's qubit is in state $|\searrow\rangle$. To calculate the result of Bob's measurement we have to rewrite this using Bob's basis. (We did a similar calculation on page 51.)

Writing everything using two-dimensional kets, we have:

$$|\searrow\rangle = \begin{bmatrix} 1/2 \\ -\sqrt{3}/2 \end{bmatrix} \qquad |\swarrow\rangle = \begin{bmatrix} -1/2 \\ -\sqrt{3}/2 \end{bmatrix} \qquad |\nearrow\rangle = \begin{bmatrix} \sqrt{3}/2 \\ -1/2 \end{bmatrix}$$

We multiply $|\searrow\rangle$ by the matrix with rows given by the bras of Bob's basis.

$$\begin{bmatrix} -1/2 & -\sqrt{3}/2 \\ \frac{\sqrt{3}}{2} & -1/2 \end{bmatrix} \begin{bmatrix} 1/2 \\ -\sqrt{3}/2 \end{bmatrix} = \begin{bmatrix} 1/2 \\ \sqrt{3}/2 \end{bmatrix}$$

This tells us that $|\searrow\rangle = \frac{1}{2}|\swarrow\rangle + \frac{\sqrt{3}}{2}|\nearrow\rangle$. When Bob makes his measurement, he will get 0 with probability 1/4 and 1 with probability 3/4. So, when Alice gets 0, Bob will get 0 with probability 1/4. It is easy to check the other case. If Alice gets 1, Bob's probability of also getting 1 is 1/4.

The other cases are all similar: If Bob and Alice measure in different directions, they will agree 1/4 of the time and disagree 3/4 of the time.

To summarize: One-third of the time they measure in the same direction and agree each time; two-thirds of the time they measure in different directions and agree on just one-quarter of these measurements. This gives the proportion of As in the string consisting of As and Ds as

$$\frac{1}{3} \times 1 + \frac{2}{3} \times \frac{1}{4} = \frac{1}{2}.$$

The conclusion is that the quantum mechanics model gives the answer that in the long run the proportion of As should be one-half.

We now look at the classical model.

The Classical Answer

The classical view is that the measurements in all directions are determined right from the start. There are three directions. A measurement in each direction can yield either a 0 or a 1. This gives us eight configurations: 000, 001, 010, 011, 100, 101, 110, 111, where the leftmost digit gives us the answer if we were to measure in the basis $(|\uparrow\rangle, |\downarrow\rangle)$, the middle digit gives us the answer if we were to measure in the basis $(|\searrow\rangle, |\nwarrow\rangle)$, and the rightmost digit gives us that answer if we were to measure in the basis $(|\swarrow\rangle, |\nearrow\rangle)$.

The entanglement just means that the configurations for Alice's and Bob's qubits are identical—if Alice's qubit has configuration 001, then so does Bob's. We now have to figure out what happens when Alice and Bob choose a direction. For example, if their electrons are in configuration 001 and Alice measures using basis $(|\uparrow\rangle, |\downarrow\rangle)$, and Bob measures using the third basis, then Alice will get a measurement of 0 and Bob a measurement of 1, and they will disagree.

The table below gives all the possibilities. The left column gives the configurations, and the top row gives the possibilities for Alice and Bob's measurement bases. We will use letters to represent the bases. We denote $(|\uparrow\rangle, |\downarrow\rangle)$ by a, $(|\searrow\rangle, |\nwarrow\rangle)$ by b, and $(|\swarrow\rangle, |\nearrow\rangle)$ by c. We will list Alice's basis first and then Bob's. So, for example, (b, c) means Alice is choosing $(|\searrow\rangle, |\nwarrow\rangle)$ and Bob is choosing $(|\swarrow\rangle, |\nearrow\rangle)$. The entries in the table show whether the measurements agree or disagree.

Config.	Measurement directions								
	(a,a)	(a,b)	(a,c)	(b,a)	(b,b)	(b,c)	(c,a)	(c,b)	(c,c)
000	A	A	A	A	A	A	A	A	A
001	A	A	D	A	A	D	D	D	A
010	A	D	A	D	A	D	A	D	A
011	A	D	D	D	A	A	D	A	A
100	A	D	D	D	A	A	D	A	A
101	A	D	A	D	A	D	A	D	A
110	A	A	D	A	A	D	D	D	A
111	A	A	A	A	A	A	A	A	A

We do not know the probabilities that should be assigned to the configurations. There are eight possible configurations, so it might seem plausible that they each occur with probability 1/8, but they perhaps are not all equal. Our mathematical analysis will make no assumption about these probabilities' values. We can, however, assign definite probabilities to the measurement directions. Both Bob and Alice are choosing each of their three bases with equal probability, so each of the nine possible pairs of bases occurs with probability 1/9.

Notice that each row contains at least five As, telling us that given a pair of qubits with any configuration the probability of getting an A is at least 5/9. Since the probability of getting an A is at least 5/9 for each of the spin configurations, we can deduce that the probability overall must be at least 5/9, no matter what proportion of time we get any one configuration.

We have now derived Bell's result. The quantum theory model tells us that Alice's and Bob's sequences will agree exactly half the time. The classical model tells us that Alice's and Bob's sequences will agree at least 5/9ths of the time. It gives us a test to distinguish between the two theories.

Bell published his inequality in 1964. Sadly, this was after the death of both Einstein and Bohr, so neither ever realized that there would be an experimental way of deciding their debate.

Actually carrying out the experiment is tricky. John Clauser and Stuart Freedman first performed it in 1972. It showed that the quantum mechanical predictions were correct. The experimenters, however, had to make some assumptions that could not be checked, leaving some chance that the classical view could still be correct. The experiment has since been repeated with increasing sophistication. It has always been in agreement with quantum mechanics, and there seems little doubt now that the classical model is wrong.

There were three potential problems with the earliest experiments. The first was that Alice and Bob were too close to one another. The second was that their measurements were missing too many entangled particles. The third was that Alice's and Bob's choices of measurement direction were not really random. If the experimenters are close to one another, it is theoretically possible that the measurements could be influenced by some other mechanism. For example, as soon as the first measurement is made, a photon travels to influence the second measurement. To ensure that this is not occurring, the measurers need to be far enough apart to know that the time interval between their measurements is less than the time it takes for a photon to travel between them. To counter this loophole, entangled photons are used. Unlike entangled electrons, entangled photons can travel long distances without interacting with the outside world.

Unfortunately, this property of not interacting readily with the outside world makes it difficult to measure them. In experiments involving photons, many of the entangled photons escape measurement, so it is theoretically possible that there is some selection bias going on—the results are reflecting the properties of a nonrepresentative sample. To counter the selection bias loophole, electrons have been used. But if electrons are used, how do you get the entangled electrons far enough apart before you measure them?

This is exactly the problem that the team from Delft, which we mentioned in the previous chapter, solved using electrons trapped in diamonds that are entangled with photons. Their experiment seems to have closed both loopholes simultaneously.*

The problem of randomness is harder. If the Copenhagen interpretation is correct, producing streams of random numbers is easy. If we are questioning this interpretation as it relates to randomness, however, we need to test strings of numbers and see if they are random. There are many tests to look for underlying patterns among the numbers. These tests, unfortunately, can prove only a negative. If a string fails a test, then we know the string is not random. Passing the test is a good sign, but it is not proof that the string is

* The paper "Loophole-free Bell inequality violation using electron spins separated by 1.3 kilometres" by B. Hensen et al. was published in *Nature* in 2015.

random. All we can say is that no quantum mechanical generated string has failed a test for randomness.

Clever ways have been chosen to ensure that the direction Alice chooses to measure is not correlated to Bob's. But again, it is not possible to rule out the possibility that some hidden variable theory determines what, we think, are uncorrelated random outcomes.

Most people consider that Einstein has been proved wrong, but that his theory made sense. Bell, in particular, believed that the classical theory was the better of the two theories up until he saw the results of the experiments, saying, "This is so rational that I think that when Einstein saw that, and the others refused to see it, he was the rational man. The other people, although history has justified them, were burying their heads in the sand. … So for me, it is a pity that Einstein's idea doesn't work. The reasonable thing just doesn't work."**

I am in total agreement with Bell. When you first meet these ideas, it seems to me that Einstein's view is the natural one to take. I am surprised that Bohr was so convinced that it was wrong. Bell's result, often called Bell's theorem, resulted in Bell's being nominated for the Nobel Prize in physics. Many people think that if he hadn't died of a stroke at the relatively young age of sixty-one he would have received it. Interestingly, there is a street in Belfast named after Bell's theorem—this might be the only theorem that you can enter into Google Maps and get a location.

We have to abandon the standard assumption of local reality. When particles are entangled, but perhaps far apart, we should not think of spin as a local property associated with each of the particles separately; it is a global property that has to be considered in terms of the pair of particles.

Before we leave our discussion of quantum mechanics, we should also look at one other unusual aspect of the theory.

Measurement

In our description of quantum mechanics we describe a state vector as jumping to a basis vector when we make a measurement. Everything is deterministic until we make a measurement, and then it jumps to one of the basis vectors. The probabilities for jumping to each of the basis vectors

** J. Bernstein, *Quantum Profiles* (Princeton: Princeton University Press, 1991), 84.

are known exactly, but they are still probabilities. The theory changes from being deterministic to probabilistic when we make a measurement.

In the general theory of quantum mechanics it is the solution of the Schrödinger's wave equation that collapses when a measurement is made. Erwin Schrödinger, of the eponymous equation, was very uncomfortable with this idea of waves collapsing to states given by probabilities.

A significant problem is that what we mean by measurement is not defined. It is not part of quantum mechanics. Measurements cause jumps, but what do we mean by measurement? Sometimes the word *observation* has been used instead of *measurement*, and this has led some people to talk about consciousness causing the jump, but this seems unlikely. The standard explanation is that the measurement involves an interaction with a macroscopic device. The measuring device is large enough that it can be described using classical physics and does not have to be incorporated into the quantum theoretical analysis—that whenever we make a measurement we have to interact physically with the object being measured, and this interaction causes the jump. But this explanation is not entirely satisfactory. It seems a plausible description, but it lacks mathematical precision.

Various interpretations of quantum mechanics have been proposed, each trying to eliminate something that seems problematic in the Copenhagen interpretation.

The *many-worlds* interpretation deals with the measurement problem by saying that it only appears that the state vector jumps to one of the possibilities, but in fact there are different universes and each of the possibilities is an actual occurrence in one of the many universes. The version of you in this universe sees one outcome, but there are other versions of you in other universes that see the other outcomes.

Bohmian mechanics tackles the introduction of probabilities. It is a deterministic theory in which particles behave like classical particles, but there is also a new entity called the pilot wave that gives the nonlocality properties.

There are many ardent believers in each of these theories. For example, David Deutsch, whom we will meet later, believes in the many-worlds view. But at the moment there are no scientific tests that have shown that one set of beliefs is preferable to another, unlike the local hidden-variable theory that the Bell's inequality experiments have shown to be wrong. All of the interpretations are consistent with our mathematical theory. Each interpretation is a way of trying to explain how the mathematical theory relates to reality. Perhaps, at some point there will be an insightful genius like Bell who can show that the different interpretations lead to different conclusions that can be experimentally differentiated, and that experiments will then give us some reason for choosing one interpretation over another. But at this point, most physicists subscribe to the Copenhagen interpretation. There is no convincing reason not to use this interpretation, so we shall use it without further comment from now on.

The final topic of this chapter shows that Bell's theorem is not just of academic interest. It can actually be used to give a secure way of sharing a key to be used in cryptography.

The Ekert Protocol for Quantum Key Distribution

In 1991, Artur Ekert proposed a method based on entangled qubits used in Bell's test. There are many slight variations. We will present a version that uses our presentation of Bell's result.

Alice and Bob receive a stream of qubits. For each pair, Alice receives one and Bob receives the other. The spin states are entangled. They are always in the state $\frac{1}{\sqrt{2}}|\uparrow\rangle|\uparrow\rangle + \frac{1}{\sqrt{2}}|\downarrow\rangle|\downarrow\rangle$.

If Alice and Bob measure their respective qubit using the same orthonormal basis, then we saw that they will get either 0 or 1 with equal probability, but they will both get exactly the same answer.

We could imagine a protocol where Alice and Bob both decide to measure their qubits in the standard basis every time. They will end up with exactly the same string of bits, and the string will be a random sequence of 0s and 1s, which seems like a great way of both choosing and communicating a key. The problem, of course, is that it is not the least bit secure. If Eve is intercepting Bob's qubits, she can measure them in the standard basis and then send the resulting unentangled qubit on to Bob. The result is that Alice, Bob, and Eve all end up with identical strings of bits.

The solution is to measure the qubits using a random choice of three bases—exactly as we did with the Bell test. As in the BB84 protocol, for each measurement Alice and Bob write down both the result and the basis that they chose. After they have made $3n$ measurements, they compare the sequences of bases that they chose. This can be done on an insecure channel—they are only revealing the basis, not the result. They will agree on approximately n of them. In each place they have chosen the same basis they will have made the same measurement. They will either both have 0, or both have 1. This gives them a string of n 0s and 1s. This will be their key if Eve is not listening in.

They now test for Eve. If Eve is eavesdropping, she will have to make measurements. Whenever she does, the entangled states become unentangled. Alice and Bob look at the strings of 0s and 1s that come from the times when they chose different bases. This gives two strings of 0s and 1s with length about $2n$. From the Bell inequality calculation, they know that if their states are entangled, in each place they should only agree 1/4 of the time. However, if Eve is measuring one of the qubits the proportion of times they agree changes. For example, if Eve measures a qubit before Alice and Bob have made their measurements, it is fairly straightforward to check all the possibilities to show that the proportion of times that Alice and Bob will agree increases to 3/8. This gives them a test for the presence of Eve. They calculate the proportion of agreement. If it is 1/4, they can conclude that nobody has interfered and use the key.

The Ekert protocol has the useful feature that the process generates the key. No digits need to be generated and stored beforehand, thus eliminating one of the main security threats to encryption. This protocol has been successfully carried out in the lab using entangled photons.

Having concluded the introduction to quantum concepts, the next topic to introduce is classical computation. This is the topic of the next chapter.

6 Classical Logic, Gates, and Circuits

In this chapter we briefly study classical computation, presenting the ideas in roughly chronological order. We start with boolean functions and logic, first introduced by George Boole in the late nineteenth century. In the 1930s, Claude Shannon studied boolean algebra and realized that boolean functions could be described using electrical switches. The electrical components that correspond to boolean functions are called logic gates. Composing boolean functions becomes the study of circuits involving these gates. We will begin by studying boolean functions in terms of logic; then we will show how to translate everything into circuits and gates. The material, up to this point, is now considered standard and is contained in every introductory computer science text. But after this we look at some ideas that are usually not part of the standard introduction.

In the 1970s, the Nobel Prize–winning physicist Richard Feynman became interested in computing and, for a few years in the early 1980s, he gave a course on computation at the California Institute of Technology. These lectures were eventually written up as *Feynman Lectures on Computation*. Feynman's interest in computation was partly due to his interaction with Edward Fredkin and Fredkin's idiosyncratic views of physics and computation. Fredkin believes that the universe is a computer, and that since the laws of physics are reversible we ought to study reversible computation and reversible gates. But even though Fredkin's overarching thesis is not widely accepted in the physics community, he is recognized for having some brilliant and unconventional ideas. One of these is the billiard ball computer. Feynman's book includes a discussion of reversible gates and shows how any computation can be performed by bouncing balls off one another.

We take Feynman's approach. It turns out that reversible gates are exactly what we need for quantum computing. The billiard ball computer led Feynman to think of particles interacting instead of balls. It was the inspiration for his work on quantum computing, but we include it here mainly because of its sheer simplicity and ingenuity.

Logic

In the late nineteenth century George Boole realized that certain parts of logic could be treated algebraically—that there were laws of logic that could be expressed in terms of algebra. We adopt the now standard way of introducing boolean logic by using truth tables for the three basic operations *not, and,* and *or.*

Negation

If a statement is true, then its negation is false, and conversely, if a statement is false, then its negation is true. For example, the statement $2 + 2 = 4$ is true, and its negation $2 + 2 \neq 4$ is false. Instead of giving concrete examples of statements, we often let the symbols P, Q, and R stand in for them. So, for example, $2 + 2 = 4$ might be represented by P. The symbol $\neg$ stands for *not*; if P represents the statement $2 + 2 = 4$, then $\neg P$ stands for $2 + 2 \neq 4$. We can then summarize the basic properties of negation using our symbols: If P is true, then $\neg P$ is false. If P is false, then $\neg P$ is true.

To make things even more concise we can use the symbols T and F to denote *true* and *false* respectively. We can then define the properties using a table.

P	$\neg P$
T	F
F	T

And

The symbol for *and* is $\wedge$. If we have two statements P and Q, we can combine them to form $P \wedge Q$. The statement $P \wedge Q$ is true if and only if both of the component statements P and Q are true. We define *and* by the following table, where the first two columns give the possibilities for the truth-values of P and Q and the third column gives us the corresponding truth-value of $P \wedge Q$.

P	Q	$P \wedge Q$
T	T	T
T	F	F
F	T	F
F	F	F

Or

The symbol for *or* is $\vee$ and is defined by the following table.

P	Q	$P \vee Q$
T	T	T
T	F	T
F	T	T
F	F	F

Notice that $P \vee Q$ is true if both P and Q are true, so $P \vee Q$ is true if either one of P or Q is true and also if both are true. This is the *or* that is used in mathematics and is sometimes called the *inclusive or*. The *exclusive or* is defined to be true if either one, but not both, of P and Q is true. It is false if they are both false, but is also false if they are both true. The *exclusive or* is denoted by $\oplus$. Its truth table is below.

P	Q	$P \oplus Q$
T	T	F
T	F	T
F	T	T
F	F	F

(Later we will see why the symbol for the *exclusive or* resembles a plus sign— it corresponds to addition modulo two.)

Boolean Algebra

We start by showing how to construct the truth table for any binary expression. For concreteness we will construct the truth table of $\neg(\neg P \wedge \neg Q)$. This is done in several steps. First we write down the table for the possibilities for P and Q.

P	Q
T	T
T	F
F	T
F	F

Then we attach columns for $\neg P$ and $\neg Q$, writing in the appropriate truth-values in each case.

P	Q	$\neg P$	$\neg Q$
T	T	F	F
T	F	F	T
F	T	T	F
F	F	T	T

Next we add a column for $\neg P \wedge \neg Q$. This is true only in the case when both $\neg P$ and $\neg Q$ are true.

P	Q	$\neg P$	$\neg Q$	$\neg P \wedge \neg Q$
T	T	F	F	F
T	F	F	T	F
F	T	T	F	F
F	F	T	T	T

Finally, we get to the column associated with $\neg(\neg P \wedge \neg Q)$. This statement is true if and only if $\neg P \wedge \neg Q$ is false.

P	Q	$\neg P$	$\neg Q$	$\neg P \wedge \neg Q$	$\neg(\neg P \wedge \neg Q)$
T	T	F	F	F	T
T	F	F	T	F	T
F	T	T	F	F	T
F	F	T	T	T	F

Omitting the intermediate columns corresponding to the intermediate steps gives the following table.

P	Q	$\neg(\neg P \wedge \neg Q)$
T	T	T
T	F	T
F	T	T
F	F	F

Logical Equivalence

Notice that the truth-values in the table for $\neg(\neg P \wedge \neg Q)$ are identical to the truth-values in the table for $P \vee Q$. They have exactly the same truth-values in every case. We say that the statements $P \vee Q$ and $\neg(\neg P \wedge \neg Q)$ are *logically equivalent*. We write:

$$P \vee Q \equiv \neg(\neg P \wedge \neg Q)$$

This means that we need never use *or*. Every case where *or* occurs can be replaced using expressions involving $\neg$ and $\wedge$.

What about the *exclusive or*, which we write with $\oplus$? Can we replace this by an expression involving only the use of $\neg$ and $\wedge$? We can, and we will now show how to do this.

We consider the truth table for $\oplus$.

P	Q	$P \oplus Q$
T	T	F
T	F	T
F	T	T
F	F	F

We look for entries of T in the third column. The first occurs when P has value T and Q has value F. An expression that gives us a value of T only for those particular truth-values of P and Q is $P \wedge \neg Q$.

The next value of T in the third column occurs when P has value F and Q has value T. An expression that gives us a value of T only for those particular truth-values of P and Q is $\neg P \wedge Q$.

These are the only places where T occurs in the third column. To get an expression equivalent to the one that we want, we now join all the expressions we have generated so far using $\vee$ s, so

$$P \oplus Q \equiv (P \wedge \neg Q) \vee (\neg P \wedge Q).$$

We know

$$P \vee Q \equiv \neg(\neg P \wedge \neg Q).$$

Using this to replace $\vee$ gives

$$P \oplus Q \equiv \neg(\neg(P \wedge \neg Q) \wedge (\neg(\neg P \wedge Q))).$$

Again, this means that we that we need never use $\oplus$. Every case where $\oplus$ occurs can be replaced using expressions involving $\neg$ and $\wedge$. The method we have just used for replacing $\oplus$ using $\neg$ and $\wedge$ works quite generally.

Functional Completeness

We can think of the logical operators that we have introduced as functions. For example, $\wedge$ is a function that has two inputs, P and Q, and gives us one output; $\neg$ has one input and one output.

We could invent our own function that has a number of inputs that take on values of T and F and in each of the cases gives us a value of either T or F; such a function is called a *boolean function*. To make things more concrete, we will invent a function that has three inputs that we will label P, Q, and R. We call our function $f(P,Q,R)$. To define our function, we have to complete the third column in the following table.

P	Q	R	$f(P,Q,R)$
T	T	T	
T	T	F	
T	F	T	
T	F	F	
F	T	T	
F	T	F	
F	F	T	
F	F	F	

There are eight values that need to be filled in. There are two choices for each value, giving us a total of 2^8 possible functions. We will show that no matter how we choose our function, we can find an equivalent expression that uses only the functions $\neg$ and $\wedge$.

We use exactly the same method that we used to show that

$P \oplus Q \equiv (P \wedge \neg Q) \vee (\neg P \wedge Q).$

We begin by looking for values of T in the last column. To help make things easier to follow we will use the specific function given by the following table, but the method we use will work for any boolean function.

P	Q	R	$f(P,Q,R)$
T	T	T	F
T	T	F	F
T	F	T	T
T	F	F	F
F	T	T	F
F	T	F	T
F	F	T	F
F	F	F	T

The first T occurs when P and R have values T, and Q has value F. A function that gives us a value of T for only this set of truth-values is $P \wedge \neg Q \wedge R$. The next T occurs when P and R have values F and Q has value T. A function that gives us a value of T for only this set of truth-values is $\neg P \wedge Q \wedge \neg R$. The final T occurs when P, Q, and R all have value F. A function that gives us a value of T for only this set of truth-values is $\neg P \wedge \neg Q \wedge \neg R$.

An expression that takes on value T in just these three cases is

$$(P \wedge \neg Q \wedge R) \vee (\neg P \wedge Q \wedge \neg R) \vee (\neg P \wedge \neg Q \wedge \neg R),$$

so

$$f(P,Q,R) \equiv (P \wedge \neg Q \wedge R) \vee (\neg P \wedge Q \wedge \neg R) \vee (\neg P \wedge \neg Q \wedge \neg R).$$

The final step is to replace $\vee$ using the fact that

$$P \vee Q \equiv \neg(\neg P \wedge \neg Q).$$

Replacing the first occurrence gives

$$f(P,Q,R) \equiv \neg(\neg(P \wedge \neg Q \wedge R) \wedge \neg(\neg P \wedge Q \wedge \neg R)) \vee (\neg P \wedge \neg Q \wedge \neg R).$$

Replacing the second occurrence tells us that $f(P,Q,R)$ is logically equivalent to

$$\neg(\neg[\neg(\neg(P \wedge \neg Q \wedge R) \wedge \neg(\neg P \wedge Q \wedge \neg R))] \wedge \neg[\neg P \wedge \neg Q \wedge \neg R]).$$

This method works in general. If f is a function that is defined by a truth table, then f is logically equivalent to some expression that involves only the functions $\neg$ and $\wedge$. Since we can generate any boolean function whatsoever using just these two functions, we say that $\{\neg, \wedge\}$ is a *functionally complete* set of boolean operators.

It might seem surprising that we can generate any function defined by a truth table using just $\neg$ and $\wedge$, but incredibly, we can do even better. There is a binary operator called *Nand*, and any boolean function is logically equivalent to some expression that only uses the *Nand* operator.

Nand

Nand is a portmanteau word formed from combining *not* and *and*. It is denoted by $\uparrow$. It can be defined by

$$P \uparrow Q = \neg(P \wedge Q),$$

or, equivalently, by the following truth table:

P	Q	$P \uparrow Q$
T	T	F
T	F	T
F	T	T
F	F	T

We know that $\{\neg, \wedge\}$ is a functionally complete set of operators, so to show that *Nand* by itself is functionally complete—that any boolean operator can be rewritten as an equivalent function that just uses *Nand* — we just need to show that both *and* and *not* have equivalent expressions that are written solely in terms of *Nand*.

Consider the following truth table, which considers just the statement P, then $P \wedge P$ and finally $\neg(P \wedge P)$.

P	$P \wedge P$	$\neg(P \wedge P)$
T	T	F
F	F	T

Notice that the final column has the same truth-values as $\neg P$, telling us

$$\neg(P \wedge P) \equiv \neg P,$$

but $\neg(P \wedge P)$ is just $P \uparrow P$, so

$$P \uparrow P \equiv \neg P.$$

This shows that we can replace all occurrences of *not* with *Nand*. We now turn our attention to *and*.

Observe that

$$P \wedge Q \equiv \neg\neg(P \wedge Q).$$

Now, $\neg(P \wedge Q) \equiv P \uparrow Q$, so

$$P \wedge Q \equiv \neg(P \uparrow Q).$$

We can now replace *not* using the preceding identity to obtain

$$P \wedge Q \equiv (P \uparrow Q) \uparrow (P \uparrow Q).$$

Henry M. Sheffer, in 1913, first published the fact that *Nand* by itself is functionally complete. Charles Sanders Peirce also knew this fact in the late nineteenth century, but like much of his highly original work it remained unpublished until much later. (Sheffer used the symbol | for *Nand*. Many authors use, or have used, Sheffer's symbol instead of $\uparrow$. It is called the *Sheffer stroke*.)

Boolean variables take on one of two values. We have been using T and F for these, but we can use any two symbols. In particular, we can use 0 and 1. The advantage of replacing T and F by 0 and 1 is that we can then think of boolean functions as operating on bits. This is what we will do from now on.

There are two choices for how we could do the substitution. The convention is that 0 replaces F and 1 replaces T, and this what we shall use. Notice that conventionally we list T before F, but 0 before 1. Consequently, truth tables written in terms of 0 and 1 reverse the order of the rows when written in terms of T and F. This shouldn't cause any confusion, but just to hammer home the point, here are the two tables for $P \vee Q$.

P	Q	$P \vee Q$
T	T	T
T	F	T
F	T	T
F	F	F

P	Q	$P \vee Q$
0	0	0
0	1	1
1	0	1
1	1	1

Gates

Various people realized that if logic could be expressed in terms of algebra, then machines could be designed to perform logical operations, but the most influential by far was Claude Shannon, who showed that all of boolean algebra could be performed using electrical switches. This is one of the fundamental ideas underlying the circuit design of all modern computers. Remarkably, he did this while still a master's student at MIT.

At discrete time intervals either a pulse of electricity is transmitted or it is not. If at the appropriate time interval we receive a pulse of electricity, then we think of this as representing the truth-value T or, equivalently, bit value 1. If at the appropriate time interval we do not receive a pulse of electricity, then we think of this as representing the truth-value F or, equivalently, bit value 0.

The combinations of switches that correspond to our binary operators are called *gates*. The common gates have special diagrams associated to them. We look at some of these.

The *NOT* Gate

Figure 6.1 shows the symbol for the *NOT* gate. This can be thought of as a wire entering from the left and leaving from the right. If we input 1, we get output 0. If we input 0, we get output 1.

The *AND* Gate

Figure 6.2 shows the symbol for the *AND* gate. Again, it is read from left to right. It has two inputs that can be either 0 or 1 and one output. Figure 6.3 shows the four cases.

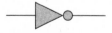

Figure 6.1
The *NOT* gate.

Figure 6.2
The *AND* gate.

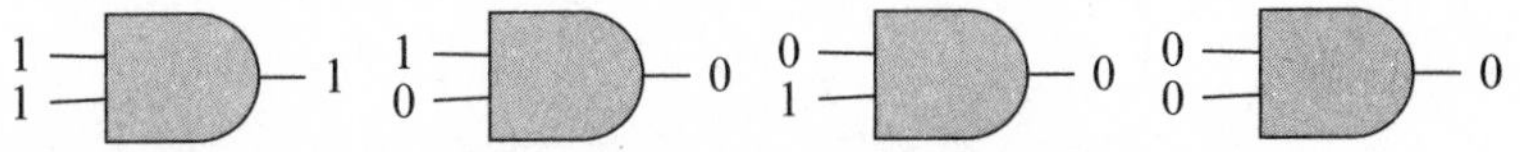

Figure 6.3
The four possibilities for inputs to the *AND* gate.

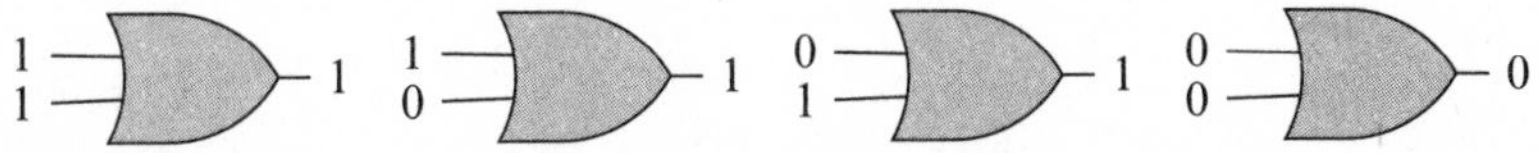

Figure 6.4
The *OR* gate.

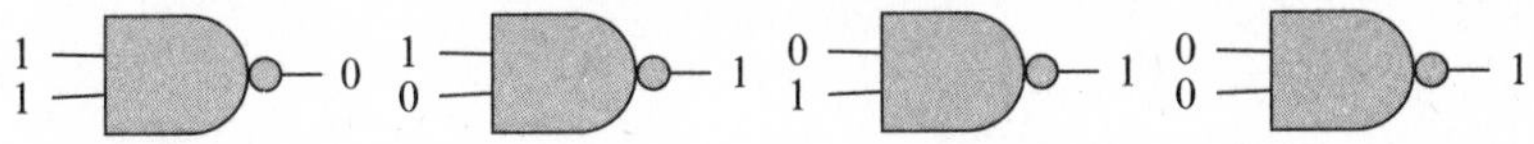

Figure 6.5
The *NAND* gate.

The *OR* Gate

Figure 6.4 shows the symbol for the *OR* gate, along with the inputs and output for the four cases.

The *NAND* Gate

Figure 6.5 shows the symbol for the *NAND* gate, along with the inputs and output for the four cases.

Circuits

We can connect the gates together to form a circuit. Despite the name, there is nothing circular about circuits. They are linear and are read from left to right. We input our bits into the wires on the left and read the output from the wires on the right. We will look at examples that correspond to the boolean functions that we looked at earlier.

We start with the boolean expression $\neg(\neg P \wedge \neg Q)$. The corresponding circuit can be given using gates. This is shown in figure 6.6, where the wires entering and leaving the gates have been labeled with the appropriate expressions. Recall that $P \vee Q \equiv \neg(\neg P \wedge \neg Q)$, so the circuit in figure 6.6 is equivalent to the *OR* gate.

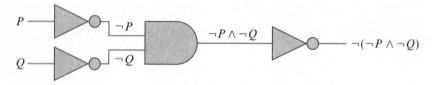

Figure 6.6
A circuit for $\neg(\neg P \wedge \neg Q)$.

Figure 6.7
A circuit for $P \uparrow P$.

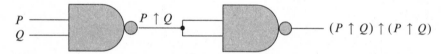

Figure 6.8
A circuit for $(P \uparrow Q) \uparrow (P \uparrow Q)$.

Our next example is $P \uparrow P$. We want to enter the same value, P, into both the inputs of our *NAND* gate. Splitting the input signal into two by connecting an additional wire achieves this. This process of splitting a signal into multiple copies is called *fan-out*. Figure 6.7 shows the circuit.

We know that $P \uparrow P \equiv \neg P$, so the circuit in figure 6.7 is equivalent to the *NOT* gate.

Our final example is the binary expression $(P \uparrow Q) \uparrow (P \uparrow Q)$. To get the two copies of $P \uparrow Q$, we again need to use fan-out. Figure 6.8 shows the circuit.

We know that $P \wedge Q \equiv (P \uparrow Q) \uparrow (P \uparrow Q)$, so the circuit in figure 6.8 is equivalent to the *AND* gate.

NAND Is a Universal Gate

Earlier we showed that the boolean function *Nand* was functionally complete. In this section we repeat the argument using gates.

Our argument started by showing that we could replace any occurrence of *or* by using the identity

$$P \vee Q \equiv \neg(\neg P \wedge \neg Q).$$

The corresponding circuit, shown in figure 6.6, shows that we need never use the *OR* gate.

The argument continued by showing that any boolean function could be constructed using combinations of *not* and *and*. Consequently, we can construct a circuit that computes any boolean function using just *NOT* and *AND* gates.

Then we showed that both *not* and *and* could be generated by *Nand* showing that *Nand* by itself was functionally complete. The analogous statement is true for the *NAND* gate. You can implement any boolean function using a circuit that just uses *NAND* gates. Instead of using the term *functionally complete*, the standard term for gates *is universal*, so *NAND* is a universal gate. But let's look at this in a little more detail.

The circuits in figures 6.7 and 6.8 show how to get rid of *NOT* and *AND* gates, replacing them with *NAND* gates. But notice that we also have to use fan-out. This operation takes one bit of information and outputs two output bits that are identical to the input bit. It might seem obvious that we can do this; it just requires connecting one piece of wire to another, but we will see later that we cannot perform this operation when it comes to quantum bits.

Gates and Computation

Gates are the fundamental building blocks of the modern computer. In addition to performing logical operations we can use gates to compute. We won't show how this can be done. (The interested reader should see the wonderful book *Code* by Charles Petzold, where he starts with switches and shows how to construct a computer.) But we will give an example to help illustrate how the ideas underlying addition can be implemented.

Recall the *exclusive or*, denoted $\oplus$. It's defined by:

$$0 \oplus 0 = 0, \qquad 0 \oplus 1 = 1, \qquad 1 \oplus 0 = 1, \qquad 1 \oplus 1 = 0.$$

This can be compared to adding odd and even whole numbers. We know:

even + even = even,   even + odd = odd,   odd + even = odd,   odd + odd = even.

Figure 6.9
The *XOR* gate.

This addition of "oddness" and "evenness" is often called *addition modulo* 2. If we let 0 stand for "even" and 1 stand for "odd," addition modulo 2 is given by ⊕. This is why the symbol contains a plus sign. (It is often easier to calculate with ⊕ thinking of addition, rather than use the exclusive or.)

The *exclusive or* gate is called *XOR* and is denoted by the symbol shown in figure 6.9.

We will use this gate to construct what is called a *half-adder*. This is a circuit that adds two binary digits. To understand what is going on we will compare it to a decimal half-adder. If we have two digits that sum to less than ten, then we just add them. So, for example, 2 + 4 = 6, 3 + 5 = 8.

If the digits sum to more than ten, however, we write down the appropriate digit, but we must remember to carry one for the next step in the computation. So, for example, 7 + 5 = 2, and we have a carry of 1.

A binary half-adder does the analogous computation. We can construct it using an *XOR* gate and an *AND* gate. The *XOR* gate computes the digit part, and the *AND* gate computes the carry.

0 + 0 = 0, with carry = 0;
0 + 1 = 1, with carry = 0;
1 + 0 = 1, with carry = 0;
1 + 1 = 0, with carry = 1.

A circuit that performs this is shown in figure 6.10. (In this picture the crossings of the wires that have dots indicate fan-out operations. The crossings without dots mean that the wires cross one another, but are not connected.)

The reason that this is called a half-adder, and not just an adder, is that it doesn't take into account that we might have a carry coming in from the step before. We look at an example where we are adding standard decimal numbers. Suppose that the calculation is to add the following four-digit numbers, where the stars represent unknown digits.

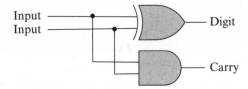

Figure 6.10
A half-adder circuit.

      **6*

+   **5*
      _____

To add the 6 and 5 we might get a digit of 1 and a carry of 1, but it is possible that we might have a carry of 1 from the first step of the calculation, in which case, the digit would be 2 and the carry 1. A full adder takes into account the possibility of an incoming carry from the step before.

We won't draw the circuit for a full binary adder, but it can be done. Since all of our gates can be replaced with *NAND* gates, we can build an adder just using *NAND* gates and fan-outs. Indeed, we can build a whole computer just using these two components.

Memory

We have shown how to use gates for logic and indicated how we can use gates to do arithmetic, but to build a computer we also need to be able to store data. This can also be done using gates. It will take us too far afield to describe in detail how to do this, but the key idea is to build a *flip-flop*. These can be built out of gates using feedback. The outputs of the gates are fed back into inputs. An example using two *NAND* gates is shown in figure 6.11. We won't describe how to implement these, but we will end by commenting that once we start using feedback it is important to get the timing of inputs and outputs exactly right. This is where the clock comes in, sending pulses of electricity at constant time intervals.

Reversible Computation

Now that we have given some idea of how a computer can be built from classical gates, we begin our study of reversible gates.

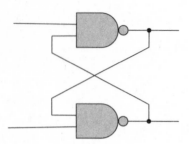

Figure 6.11
A flip-flop using two *NAND* gates.

Gates can be considered as boolean functions. For example, the *AND* gate takes two boolean inputs and gives a boolean output. Often the easiest way of representing this is through a table. (This table is exactly the same as what we have been calling a truth table.)

AND

Input		Output
0	0	0
0	1	0
1	0	0
1	1	1

We can also represent the half-adder using a table. This time there are two inputs and two outputs.

Half-adder

Input		Output	
		digit	carry
0	0	0	0
0	1	1	0
1	0	1	0
1	1	0	1

In this section we will look at reversible gates. These correspond to invertible functions. Given an output, can we determine what the input was? If we can in every case, the function is invertible—the gate is reversible.

Looking at *AND*, if we get an output of 1, then we know that the input values must have been both 1, but if we get an output value of 0 there are three pairs of input values that have this output, and if we are not given any other information, we have no way of knowing which one of the three possibilities was actually input. Consequently, *AND* is not a reversible gate.

The half-adder is also not reversible. There are two pairs of input values that give a digit of 1 and a carry of 0. In both of these cases we have two bits of input, but are not getting two bits of output. We have lost some information doing the computation.

The study of reversible gates and reversible computation began by looking at the thermodynamics of computation. Shannon defined entropy for information. Entropy is also defined in thermodynamics. In fact, this is where Shannon got the idea. How closely are these two entropies related to one another? Can some of the theory of computation be expressed in terms of thermodynamics? In particular, can one talk about the minimum energy required performing a calculation? John von Neumann conjectured that when information was lost energy is expended—it dissipates as heat. Rolf Landauer proved the result and gave the minimum possible amount of energy to erase one bit of information. This amount of energy is called the *Landauer limit*.

If the computation is reversible, however, no information is lost and theoretically it can be performed with no energy loss.

We will look at three reversible gates: the *CNOT*, Toffoli, and Fredkin gates.

Controlled Not Gate

The controlled not gate or *CNOT* gate takes two inputs and gives two outputs. The first input is called the control bit. If it is 0, then it has no effect on the second bit. If the control bit is 1, it acts like the *NOT* gate on the second bit. The control bit is the first input bit and denoted by x. This bit is not changed and becomes the first output. The second output equals the second input if the control bit is 0, but it is flipped when the control bit is 1. This function is given by $f(x,y) = (x, x \oplus y)$ or, equivalently, by the following table.

CNOT

Input		Output	
x	y	x	$x \oplus y$
0	0	0	0
0	1	0	1
1	0	1	1
1	1	1	0

Notice that this operation is invertible. Given any pair of output values, there is exactly one pair of input values that corresponds to it.

We can build a circuit that performs this operation using a fan-out and an *XOR* gate. This is shown in figure 6.12.

This, however, is not the picture that is most commonly used. The usual picture is the simplified version shown in figure 6.13.

The *CNOT* gate is not just invertible, but it also has the nice property that it is its own inverse. This means that if you put two *CNOT* gates in series, where the output of the first gate becomes the input of the second gate, the output from the second gate is identical to the input to the first gate. The second gate undoes what the first gate does. To see this, we know that applying the *CNOT* gate once is given by

$$f(x,y) = (x, x \oplus y).$$

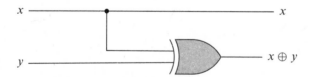

Figure 6.12
A circuit for *CNOT*.

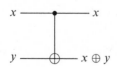

Figure 6.13
Usual representation of *CNOT* gate.

Using this output as the input of another *CNOT* gate gives

$$f(x, x \oplus y) = (x, x \oplus x \oplus y) = (x, y) .$$

Here we have used the facts that $x \oplus x = 0$ and $0 \oplus y = y$.

We started with the input (x, y) and the output after going through the gate twice is (x, y), back where we started.

The Toffoli Gate

The Toffoli gate, invented by Tommaso Toffoli, has three inputs and three outputs. The first two inputs are control bits. They flip the third bit if they are both 1, otherwise the third bit remains the same. Since this gate is like the *CNOT* gate, but has two control bits, it is sometimes called a *CCNOT* gate. The function describing what this gate does is: $T(x, y, z) = (x, y, (x \wedge y) \oplus z)$.

This can also be given in tabular form.

Toffoli gate

Input			Output		
x	y	z	x	y	$(x \wedge y) \oplus z$
0	0	0	0	0	0
0	0	1	0	0	1
0	1	0	0	1	0
0	1	1	0	1	1
1	0	0	1	0	0
1	0	1	1	0	1
1	1	0	1	1	1
1	1	1	1	1	0

The standard diagram for this gate comes from the diagram of the *CNOT* gate (figure 6.14).

We can see from the table that the Toffoli gate is invertible—each triple of output values corresponds to exactly one triple of input values. Like the *CNOT* gate, this gate also has the property that it is its own inverse.

We know that $T(x, y, z) = (x, y, (x \wedge y) \oplus z)$. Now, using the output as the new input and applying T again gives:

$$T(x, y, (x \wedge y) \oplus z) = (x, y, (x \wedge y) \oplus (x \wedge y) \oplus z) = (x, y, z).$$

Here we use the facts that $(x \wedge y) \oplus (x \wedge y) = 0$ and $0 \oplus z = z$.

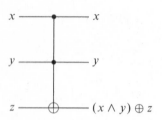

Figure 6.14
Toffoli gate.

The Toffoli gate is also universal. Recall that we can construct any boolean circuit using just *NAND* gates and fan-outs. To show that the Toffoli gate is universal, it is enough if we can show how to use it to compute both of these.

The *NAND* gate is described by $f(x,y) = \neg(x \wedge y)$, so we want a way of inputting x and y and getting an output of $\neg(x \wedge y)$. Since we are using the Toffoli gate, we will be inputting three values and getting an output of three values. Now $\neg(x \wedge y)$ is logically equivalent to $(x \wedge y) \oplus 1$. We can choose the third input value to always be 1, and we can ignore extra output values. We use

$$T(x,y,1) = (x,y,(x \wedge y) \oplus 1) = (x,y,\neg(x \wedge y))$$

to show that we can emulate the *NAND* gate by inputting x and y and reading off the third entry of the output.

We can use a similar idea for fan-out. We want to input just one value x and receive two outputs that are both x. Again, the Toffoli gate has three inputs and three outputs. We can choose the two other inputs apart from x to be fixed and as long as we get xs for two of the outputs we can ignore the third. This can be done by

$$T(x,1,0) = (x,1,x).$$

Consequently, any boolean circuit can be constructed using just Toffoli gates.

These constructions illustrate something that often arises when we use reversible gates. The number of inputs must equal the number of outputs, but often we want to compute things where the number of inputs and outputs differ. We can always do this by adding extra bits, often called ancilla bits, to the inputs, or by ignoring bits that are output. Output bits that are ignored are sometimes called garbage bits. In the example where we showed that fan-out could be done using the Toffoli gate we had $T(x,1,0) = (x,1,x)$. The 1 and 0 in the input are ancilla bits, and the 1 in the output is a garbage bit.

The Fredkin Gate

The Fredkin gate also has three inputs and three outputs. The first input is a control bit. If it is 0, the second and third inputs are unchanged. If the control bit is 1, it swaps the second and third inputs—the second output is the third input and the third output is the second input. It is defined by

$$F(0, y, z) = (0, y, z), F(1, y, z) = (1, z, y).$$

Equivalently, it is given by the following table.

Fredkin gate

Input			Output		
x	y	z	x		
0	0	0	0	0	0
0	0	1	0	0	1
0	1	0	0	1	0
0	1	1	0	1	1
1	0	0	1	0	0
1	0	1	1	1	0
1	1	0	1	0	1
1	1	1	1	1	1

It is easily seen from the table that the Fredkin gate is invertible and that, like both the CNOT and Toffoli gates, it is its own inverse. The table also has the property that the number of 1s for each input is equal to the number of 1s in the corresponding output. We will make use of this fact later when we construct a Fredkin gate using billiard balls. (When constructing billiard ball gates, you want them to have the property that the number of balls entering is equal to the number of balls leaving.) Figure 6.15 shows the diagram for this gate.

Notice that $F(0,0,1) = (0,0,1)$ and $F(1,0,1) = (1,1,0)$, so for both possible values of x,

$$F(x,0,1) = (x,x,\neg x),$$

Figure 6.15
The Fredkin gate.

telling us that we can use the Fredkin gate for both fan-out and negation. For fan-out, we think of $\neg x$ as a garbage bit. For negation, we think of both the xs as garbage bits.

If we put z equal to 0 we obtain:

$$F(0,0,0) = (0,0,0), \qquad F(0,1,0)=(0,1,0), \qquad F(1,0,0) = (1,0,0), \qquad F(1,1,0) = (1,0,1).$$

We can write this more succinctly as

$$F(x,y,0) = (x, \neg x \wedge y, x \wedge y).$$

This tells us that we can use the Fredkin gate to construct the *AND* gate (0 is an ancilla bit, and both x and $\neg x \wedge y$ are garbage bits).

Since any boolean circuit can be constructed using just *NOT* and *AND* gates along with fan-out, we can construct any boolean circuit using just Fredkin gates. Like the Toffoli gate, the Fredkin gate is universal.

We defined the Fredkin gate by

$$F(0, y, z) = (0,y,z), F(1,y,z) = (1,z,y),$$

but we will give another equivalent definition.

This gate outputs three numbers. The first number output is always equal to the first input x. The second number will be 1 if either $x = 0$ and $y = 1$ or if $x = 1$ and $z = 1$, which we can express as $(\neg x \wedge y) \vee (x \wedge z)$. The third output will be 1 if either $x = 0$ and $z = 1$ or if $x = 1$ and $y = 1$, which we can express as $(\neg x \wedge z) \vee (x \wedge y)$. Consequently, we can define this gate by

$$F(x,y,z) = (x, (\neg x \wedge y) \vee (x \wedge z), (\neg x \wedge z) \vee (x \wedge y)) \, .$$

This looks somewhat intimidating and seems much more complicated than just remembering that if $x = 0$, then both y and z are unchanged; if $x = 1$, then y and z get switched. However, there is one place where this complicated formula is useful, and that is in the next section, when we show how to construct this gate using billiard balls.

Billiard Ball Computing

We haven't discussed how to actually build gates. They can all be built from switches and wires with electric potential or its absence representing the bits 1 and 0. Fredkin showed that they could also be built using billiard balls that bounce off one another and strategically placed mirrors. A mirror is just a solid wall that the ball bounces off. (They are called mirrors because the angle of incidence is equal to the angle of reflection.) Billiard ball gates are theoretical devices; it is assumed that all collisions are totally elastic— that no energy is lost. An example of a simple gate, called the *switch gate*, is shown in figure 6.16. In these pictures the solid lines represent walls; the grid lines are drawn to help keep track of the centers of the balls.

In the picture on the left a ball has just entered via Input 1. Since we haven't entered a ball into Input 2, the ball just rolls unhindered and exits via Output 1. The picture on the right shows the analogous situation when one ball enters via Input 2 and we don't send a ball through Input 1: It rolls unhindered out of Output 2A.

There are two other possibilities for sending ball through the two input slots. Unsurprisingly, if we don't enter any balls, then no balls exit. The final and most complicated case is when balls are sent through both inputs. The assumption is that the balls have the same size, mass, speed and are entered simultaneously. Figure 6.17 indicates what happens.

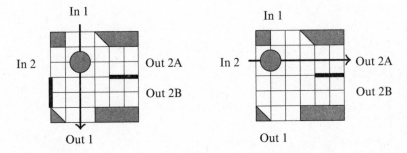

Figure 6.16
Billiard ball switch gate.

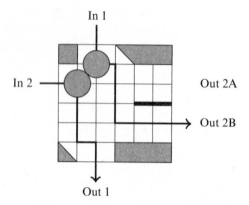

Figure 6.17
Two balls entering switch gate.

First the balls collide with one another, then they both bounce off the diagonal walls (or mirrors), then they collide again. Finally, they exit. One leaves via Output 1 and the other by Output 2 B. (The paths of the centers of the balls are indicated by the bold arrows.)

We can denote the presence of a ball by 1 and the absence by 0, and then we can summarize what the gate does in a table.

Switch gate

Input		Output		
1	2	1	2A	2B
0	0	0	0	0
0	1	0	1	0
1	0	1	0	0
1	1	1	0	1

We can construct a table with the same values using the statements x, y, $\neg x \wedge y$, and $x \wedge y$.

x	y	x	$\neg x \wedge y$	$x \wedge y$
0	0	0	0	0
0	1	0	1	0
1	0	1	0	0
1	1	1	0	1

This enables us to depict the switch as a black box with the inputs and outputs appropriately labeled, as is depicted in figure 6.18.

This picture tells us where balls enter and leave the gate. If a ball enters via x, a ball must leave via x. If a ball enters via y, a ball will leave via the $\neg x \wedge y$ exit if there is no ball entering via x and will leave via the $x \wedge y$ exit if there is also a ball entering via x. At this point you might be slightly worried by the fact that in the case when two balls enter, the balls get switched because the ball that exits via x is the ball that entered via y and the ball that exits from $x \wedge y$ is the one that entered from x. But this is not a problem. We regard the balls as being indistinguishable—we just keep track of where there are balls, not where the balls originally came from.

We can also reverse the gate as is depicted in figure 6.19. We have to be slightly careful interpreting this. If a ball enters via $\neg x \wedge y$, then there won't be a ball entering via x, and so the ball sails directly across. If a ball enters

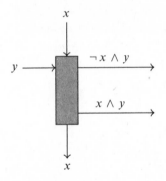

Figure 6.18
Switch gate with inputs and outputs labeled.

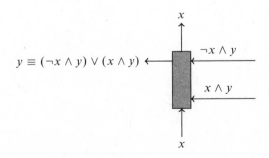

Figure 6.19
Switch gate with inputs and outputs interchanged.

via $x \wedge y$, then there will be a ball entering via x, and consequently they will collide. One ball exits through the top of the gate and one exits via the output on the left. This means that a ball will exit through the left output if either $\neg x \wedge y$ or $x \wedge y$, so this exit can be labeled $(\neg x \wedge y) \vee (x \wedge y)$. However, $(\neg x \wedge y) \vee (x \wedge y)$ is logically equivalent to y, which means that reversing the gate just reverses the arrows but leaves all the labels the same.

We are now in a position to construct a Fredkin gate. Recall that

$$F(x,y,z) = (x, (\neg x \wedge y) \vee (x \wedge z), (\neg x \wedge z) \vee (x \wedge y)).$$

We need a construction that inputs x, y and z and outputs x, $(\neg x \wedge y) \vee (x \wedge z)$ and $(\neg x \wedge z) \vee (x \wedge y))$. This can be done with four switch gates and a lot of ingenuity. It is depicted in figure 6.20.

In this picture, the right angles in the paths are obtained by bouncing off diagonally placed mirrors. The only other interactions occur in the switch gates. Paths crossing don't indicate collisions; the balls pass through the intersection points at different times. To make sure that balls don't collide where they shouldn't and do collide where they should, we can always add delays to paths by adding little detours to paths using mirrors. For example,

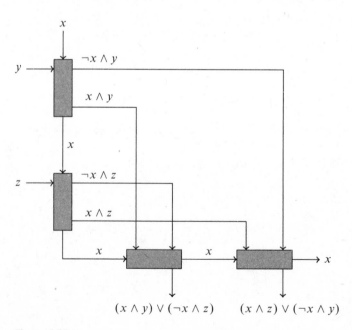

Figure 6.20
Fredkin gate constructed from switch gates.

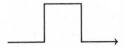

Figure 6.21
Delay added to straight-line path.

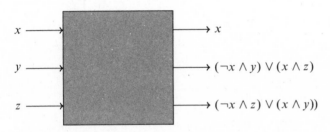

Figure 6.22
Billiard-ball Fredkin gate to be used in circuits.

we can add a little delay by changing a straight-line path to one like the one depicted in figure 6.21.

By putting mirrors in the appropriate places and adding delays, we can construct the gate so that the outputs are lined up with the inputs and when balls enter at the same time they leave at the same time. (This is depicted in figure 6.22.) We can then form circuits that contain more than one Fredkin gate.* Since the Fredkin gate is universal, it can be used to construct any boolean circuit. Consequently, any boolean circuit can be constructed using just billiard balls and mirrors.

Fredkin believes that the universe is a computer. He didn't convince Feynman of this, but the billiard ball computer did impress him. As they both realized, any slight error in the position or velocity of a ball would result in an error that would propagate and get amplified. Collisions are never perfectly elastic; there is always friction and heat is lost. The billiard ball computer is clearly just a theoretical machine, not something that can be constructed in practice. But this machine does conjure images of atoms bouncing off one another, and it led Feynman to consider gates based on quantum mechanics rather than classical mechanics. We look at this idea in the next chapter.

* There is a great animation showing this gate with balls entering and leaving on the website http://www.bubblycloud.com/billiard/fredkin-from-switches.html.

7 Quantum Gates and Circuits

Quantum gates and circuits are a natural extension of both classical gates and circuits. They are also another way of thinking about the mathematics that describes sending qubits from Alice to Bob.

I commute by train. Often the train I am on is stationary, with another train also at a standstill just inches away from my window. One train will move slowly. Sometimes it is impossible to tell whether it is my train or the other one that is moving without turning to look out of the window on the opposite side. We could be inching forward, or the train moving in the opposite direction could be inching forward. Both scenarios fit. The same analysis applies to Bob's measurements. We can either think of Bob as rotating his measuring apparatus, or we can think of Bob as keeping his apparatus in the same direction as Alice, but somehow the qubit gets rotated on the trip from Alice to Bob. When Alice and Bob are far apart, it often makes sense to think of Bob's apparatus as being rotated. But we are going to send qubits to ourselves. We could think of our apparatus rotating during the travel time, but it is more natural to think of the apparatus as fixed and the qubit as being rotated. We think of the rotation as happening between the time it is sent and the time it is measured. Sending the qubits through a quantum gate performs this rotation. Previously we said that choosing directions to measure our qubits correspond to choosing an orthogonal matrix. Now we think of the directions we are measuring as being fixed and the orthogonal matrix as corresponding to a gate that the qubits pass through. Before we look at examples, we will introduce some new names for our basis kets.

Qubits

Since we are going to think of our measuring device as being fixed, we need to use only one ordered basis for both sending and receiving qubits. The natural basis to choose is the standard one $\left(\begin{bmatrix}1\\0\end{bmatrix},\begin{bmatrix}0\\1\end{bmatrix}\right)$. Earlier we denoted this as $(|\uparrow\rangle,|\downarrow\rangle)$. But we also associated the first vector in the ordered basis to the bit 0 and the second vector to 1. Now that we are solely going to use this basis it makes sense to give our kets new names that reflect how they relate to bits. We let $|0\rangle$ denote $\begin{bmatrix}1\\0\end{bmatrix}$ and $|1\rangle$ denote $\begin{bmatrix}0\\1\end{bmatrix}$.

In general a qubit will have the form $a_0|0\rangle + a_1|1\rangle$, where $a_0^2 + a_1^2 = 1$. When we measure it, either the state jumps to $|0\rangle$ and we read 0, or the state jumps to $|1\rangle$ and we read 1. The first occurs with probability a_0^2, the second with probability a_1^2.

Usually we have a system with more than one qubit, which means that we have to form tensor products. For a system with two qubits the underlying ordered basis is

$$\left(\begin{bmatrix}1\\0\end{bmatrix}\otimes\begin{bmatrix}1\\0\end{bmatrix},\begin{bmatrix}1\\0\end{bmatrix}\otimes\begin{bmatrix}0\\1\end{bmatrix},\begin{bmatrix}0\\1\end{bmatrix}\otimes\begin{bmatrix}1\\0\end{bmatrix},\begin{bmatrix}0\\1\end{bmatrix}\otimes\begin{bmatrix}0\\1\end{bmatrix}\right).$$

This can be written as $(|0\rangle\otimes|0\rangle,|0\rangle\otimes|1\rangle,|1\rangle\otimes|0\rangle,|1\rangle\otimes|1\rangle)$. As we noted before, we often suppress the tensor product symbols, and so we write the product even more succinctly as $(|0\rangle|0\rangle,|0\rangle|1\rangle,|1\rangle|0\rangle,|1\rangle|1\rangle)$. Finally, we make the convention that we let $|ab\rangle$ denote $|a\rangle|b\rangle$, giving the representation $(|00\rangle,|01\rangle,|10\rangle,|11\rangle)$ that is short and easy to read.

How does this connect to gates? This is what we will consider next. We start by reexamining the CNOT gate.

The CNOT Gate

As we saw, the classical CNOT gate takes two input bits and gives two output bits. It's defined by the table:

CNOT

Input		Output	
x	y	x	$x \oplus y$
0	0	0	0
0	1	0	1
1	0	1	1
1	1	1	0

We extend this to qubits in the natural way—replacing 0 by $|0\rangle$, and 1 by $|1\rangle$. The table becomes:

CNOT

Input		Output	
x	y	x	$x \oplus y$
$|0\rangle$	$|0\rangle$	$|0\rangle$	$|0\rangle$
$|0\rangle$	$|1\rangle$	$|0\rangle$	$|1\rangle$
$|1\rangle$	$|0\rangle$	$|1\rangle$	$|1\rangle$
$|1\rangle$	$|1\rangle$	$|1\rangle$	$|0\rangle$

This can be written more succinctly using our compact notation for tensor products.

CNOT

Input	Output
$|00\rangle$	$|00\rangle$
$|01\rangle$	$|01\rangle$
$|10\rangle$	$|11\rangle$
$|11\rangle$	$|10\rangle$

The table tells us what happens to the basis vectors. We then extend to linear combinations of the basis vectors in the obvious way.

$$CNOT\,(r|00\rangle + s|01\rangle + t|10\rangle + u|11\rangle) = r|00\rangle + s|01\rangle + u|10\rangle + t|11\rangle$$

It just flips the probability amplitudes of $|10\rangle$ and $|11\rangle$.

We keep using the diagram we used previously for the *CNOT* gate, but we must be careful about how we interpret it. For classical bits, the bit entering the top wire on the left, leaves the top wire on the right unchanged. This is still true for qubits if the top qubit is either $|0\rangle$ or $|1\rangle$, but it is not true for other qubits.

For example, we take $\frac{1}{\sqrt{2}}|0\rangle + \frac{1}{\sqrt{2}}|1\rangle$ as the top qubit and $|0\rangle$ for the bottom one.

The input is $\left(\frac{1}{\sqrt{2}}|0\rangle + \frac{1}{\sqrt{2}}|1\rangle\right) \otimes |0\rangle = \frac{1}{\sqrt{2}}|00\rangle + \frac{1}{\sqrt{2}}|10\rangle$. This is sent by the

$CNOT$ gate to $\frac{1}{\sqrt{2}}|00\rangle + \frac{1}{\sqrt{2}}|11\rangle$.

This state, as we recognize from the EPR experiment, is an entangled state. Consequently, we cannot assign individual states to the top and bottom wires on the right side. We draw the diagram in the following way.

$$\frac{1}{\sqrt{2}}|0\rangle + \frac{1}{\sqrt{2}}|1\rangle \ \underline{\hspace{2cm}\bullet\hspace{2cm}} \Big\} \ \frac{1}{\sqrt{2}}|00\rangle + \frac{1}{\sqrt{2}}|11\rangle$$
$$|0\rangle \ \underline{\hspace{2cm}\oplus\hspace{2cm}}$$

The wires represent our electrons or photons. These are separate objects and can be far apart. We will often talk about the top qubit and the bottom qubit and think of them as being far apart. But, remember, if they are entangled, a measurement on one will affect the state of the other.

This example illustrates how we will often use this gate. We can input two unentangled qubits and use the gate to entangle them.

Quantum Gates

Notice that the $CNOT$ gate permutes the basis vectors. Permuting the basis vectors in an ordered orthonormal basis gives another ordered orthonormal basis, and we know that associated with any of these bases is an orthogonal matrix. Consequently, the matrix corresponding to the $CNOT$ gate is orthogonal. In fact, all of the reversible gates that we introduced in the last chapter permute basis vectors. They all correspond to orthogonal matrices.

This gives us the definition of quantum gates. They are just operations that can be described by orthogonal matrices.

Just as for classical computation, we want to assemble a small collection of simple gates that we can connect together to form circuits. We start by looking at the simplest gates, those that act on just one qubit.

Quantum Gates Acting on One Qubit

In classical, reversible computation there are only two possible boolean operators that act on one bit: the identity that leaves the bit unchanged, and *NOT*, which flips the values of 0 and 1. For qubits there are infinitely many possible gates!

We begin by looking at the two quantum gates that correspond to the classical identity and that both leave the qubits $|0\rangle$ and $|1\rangle$ unchanged. Then we will look at the two quantum gates corresponding to flipping the qubits $|0\rangle$ and $|1\rangle$. These four gates are named after Wolfgang Pauli and are called the *Pauli transformations*.

The Gates *I* and *Z*

The gate I is just the identity matrix $\begin{bmatrix} 1 & 0 \\ 0 & 1 \end{bmatrix}$.

We will see how I acts on an arbitrary qubit $a_0|0\rangle + a_1|1\rangle$.

$$I(a_0|0\rangle + a_1|1\rangle) = \begin{bmatrix} 1 & 0 \\ 0 & 1 \end{bmatrix}\begin{bmatrix} a_0 \\ a_1 \end{bmatrix} = \begin{bmatrix} a_0 \\ a_1 \end{bmatrix} = a_0|0\rangle + a_1|1\rangle.$$

Unsurprisingly, I acts as the identity and leaves qubits totally unchanged.

The gate Z is defined by the matrix $\begin{bmatrix} 1 & 0 \\ 0 & -1 \end{bmatrix}$.

Again, let's see how Z acts on an arbitrary qubit $a_0|0\rangle + a_1|1\rangle$.

$$Z(a_0|0\rangle + a_1|1\rangle) = \begin{bmatrix} 1 & 0 \\ 0 & -1 \end{bmatrix}\begin{bmatrix} a_0 \\ a_1 \end{bmatrix} = \begin{bmatrix} a_0 \\ -a_1 \end{bmatrix} = a_0|0\rangle - a_1|1\rangle.$$

So, Z leaves the probability amplitude of $|0\rangle$ unchanged, but it changes the sign of the probability amplitude of $|1\rangle$. But let's look at what Z does a little more carefully.

First, we will look at how it acts on the basis vectors. We have $Z(|0\rangle) = |0\rangle$ and $Z(|1\rangle) = -|1\rangle$. But recall that a state vector is equivalent to that state vector multiplied by -1, so $-|1\rangle$ is equivalent to $|1\rangle$; consequently, Z preserves both of the basis vectors, but it is not the identity. If we apply Z to the qubit $\frac{1}{\sqrt{2}}|0\rangle + \frac{1}{\sqrt{2}}|1\rangle$, we obtain

$$\frac{1}{\sqrt{2}}|0\rangle - \frac{1}{\sqrt{2}}|1\rangle,$$

and, as we showed on page 42, $\frac{1}{\sqrt{2}}|0\rangle + \frac{1}{\sqrt{2}}|1\rangle$ is distinguishable from, not equivalent to, $\frac{1}{\sqrt{2}}|0\rangle - \frac{1}{\sqrt{2}}|1\rangle$.

Even though the transformation Z preserves both the basis vectors, it changes every other qubit! This operation of changing the sign of a probability amplitude is sometimes called *changing the relative phase* of the qubit.

The Gates *X* and *Y*

The gates X and Y are given by:*
They both correspond to *NOT* in that they interchange $|0\rangle$ and $|1\rangle$. The gate X just flips, while Y flips and changes the relative phase.

$$X = \begin{bmatrix} 0 & 1 \\ 1 & 0 \end{bmatrix} \qquad Y = \begin{bmatrix} 0 & 1 \\ -1 & 0 \end{bmatrix}.$$

The Hadamard Gate

The last and most important gate that acts on one bit is the Hadamard gate, H. It is defined by

$$H = \begin{bmatrix} \frac{1}{\sqrt{2}} & \frac{1}{\sqrt{2}} \\ \frac{1}{\sqrt{2}} & -\frac{1}{\sqrt{2}} \end{bmatrix} = \frac{1}{\sqrt{2}} \begin{bmatrix} 1 & 1 \\ 1 & -1 \end{bmatrix}.$$

* Most authors define the matrix Y to be $-i$ times the matrix we have given. We have chosen not to use any complex numbers. Our choice for Y simplifies things slightly when we consider superdense coding and quantum teleportation.

This gate is often used to put the standard basis vectors into superpositions:

$$H(|0\rangle) = \frac{1}{\sqrt{2}}(|0\rangle + |1\rangle) \qquad\qquad H(|1\rangle) = \frac{1}{\sqrt{2}}(|0\rangle - |1\rangle)$$

In diagrams, gates that act on one qubit are denoted by a square with the appropriate letter drawn in the center. For example, the Hadamard gate acting on one bit is denoted by the following.

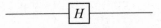

We have named five quantum gates that act on just one qubit. Of course, there are infinitely more. Any rotation will give us an orthogonal matrix, and there are infinitely many of these, all of which can be considered as gates.

Are There Universal Quantum Gates?

In classical computing, we found that every boolean function could be given by a circuit that used only Fredkin gates, telling us that the Fredkin gate is universal. We also saw that *NAND*, along with fan-out, was universal. Are there universal quantum gates?

In the classical case, there are only finitely many boolean functions with a given number of variables. There are just four boolean functions of one variable. There are sixteen of two variables. In general, there are 2^{2^n} possible functions with n variables. Things are very different with quantum gates. As we have seen there are uncountably many possible gates that can act on just one qubit. If we take a finite number of gates and connect them in a finite number of ways, we will end up with a countable number of circuits. So, it is just not possible to have a finite number of gates generate an uncountably infinite number of circuits.

The short answer to the question of whether or not there is a finite set of quantum gates that is universal is just "no." However, even though it is impossible to have a finite number of quantum gates that will generate every other possible quantum circuit, people have shown there is a finite collection of gates that can be used to *approximate* every possible circuit, but we will not go into this. All of the circuits that we need can be constructed from the gates that we have introduced; five that act on just one qubit, and one, the *CNOT* gate, that acts on two qubits.

No Cloning Theorem

We first came across the fan-out operation when we were looking at classical circuits. One input wire is connected to two output wires. The input signal is split into two identical copies.

We then looked at reversible gates. For these, if you have two outputs, then you must also have two inputs. We could get the fan-out operation by using an ancilla bit—taking the second input always to be 0. One way of doing this is with the *CNOT* gate.

CNOT $(|0\rangle|0\rangle) = |0\rangle|0\rangle$, *CNOT* $(|1\rangle|0\rangle) = |1\rangle|1\rangle$, so *CNOT* $(|x\rangle|0\rangle) = |x\rangle|x\rangle$, if $|x\rangle$ is either $|0\rangle$ or $|1\rangle$. Unfortunately, if $|x\rangle$ is not $|0\rangle$ or $|1\rangle$, we don't end up with two copies. We saw this when we input $\left(\frac{1}{\sqrt{2}}|0\rangle + \frac{1}{\sqrt{2}}|1\rangle\right)|0\rangle$ into the *CNOT* gate. It resulted in an entangled state, not two copies of the left qubit. We can use *CNOT* to copy classical bits, but not general qubits.

The term fan-out is restricted to classical computing. We use the word *cloning* for the analogous idea in quantum computing. Cloning is like fan-out, but for qubits. We want to make copies not just of classical bits but also of qubits. We want a gate that inputs a general qubit $|x\rangle$ and a fixed second input $|0\rangle$ (an ancilla bit) and outputs two copies of $|x\rangle$. A diagram of our desired gate follows.

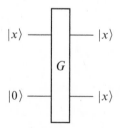

The question of cloning becomes the question of whether or not the gate G can exist. We will show that it cannot, showing that it is impossible to clone general qubits. We do this by supposing that there is such a gate and then showing that two contradictory consequences follow logically from this assumption. Since our argument is logically sound and contradictions should not occur, we conclude that our initial assumption that G existed was false. Here's the argument.

If G exists, we know that its cloning property gives:

1. $G(|0\rangle|0\rangle) = |0\rangle|0\rangle$.

2. $G(|1\rangle|0\rangle) = |1\rangle|1\rangle$.

3. $G\left(\left(\frac{1}{\sqrt{2}}|0\rangle + \frac{1}{\sqrt{2}}|1\rangle\right)|0\rangle\right) = \left(\frac{1}{\sqrt{2}}|0\rangle + \frac{1}{\sqrt{2}}|1\rangle\right)\left(\frac{1}{\sqrt{2}}|0\rangle + \frac{1}{\sqrt{2}}|1\rangle\right)$.

These three statements can be restated to give:

1. $G(|00\rangle) = |00\rangle$.

2. $G(|10\rangle) = |11\rangle$.

3. $G\left(\frac{1}{\sqrt{2}}|00\rangle + \frac{1}{\sqrt{2}}|10\rangle\right) = \frac{1}{2}(|00\rangle + |01\rangle + |10\rangle + |11\rangle)$.

The gate G, like all matrix operators, must be linear, which means that

$$G\left(\frac{1}{\sqrt{2}}|00\rangle + \frac{1}{\sqrt{2}}|10\rangle\right) = \frac{1}{\sqrt{2}}G(|00\rangle) + \frac{1}{\sqrt{2}}G(|10\rangle).$$

Replacing $G(|00\rangle)$ and $G(|10\rangle)$ using statements (1) and (2) gives

$$G\left(\frac{1}{\sqrt{2}}|00\rangle + \frac{1}{\sqrt{2}}|10\rangle\right) = \frac{1}{\sqrt{2}}|00\rangle + \frac{1}{\sqrt{2}}|11\rangle.$$

But statement (3) says that

$$G\left(\frac{1}{\sqrt{2}}|00\rangle + \frac{1}{\sqrt{2}}|10\rangle\right) = \frac{1}{2}(|00\rangle + |01\rangle + |10\rangle + |11\rangle).$$

However,

$$\frac{1}{\sqrt{2}}|00\rangle + \frac{1}{\sqrt{2}}|11\rangle \neq \frac{1}{2}(|00\rangle + |01\rangle + |10\rangle + |11\rangle).$$

So we have shown that if G exists then two things that are not equal must be equal. This is a contradiction. The only logical conclusion is that G cannot exist, and it is impossible to construct a gate that clones general qubits. The argument we have given used $|0\rangle$ for the ancilla bit. There is nothing special about this. Exactly the same argument can be used whatever value is chosen for this bit.

The inability to clone a qubit has many important consequences. We want to be able to back up files and send copies of files to other people. Copying is ubiquitous. Our everyday computers are based on von Neumann architecture, which is heavily based on the ability to copy. When we run a program we are always copying bits from one place to another. In quantum computing this is not possible for general qubits. So, if programmable quantum computers are designed they will not be based on our current architecture.

At first, the fact that we cannot clone qubits seems like a serious drawback, but there are a couple of important comments that need to be made.

Often we want to prevent copying. We want to secure our data—we don't want our communications to be tapped. Here, as we saw with Eve, the fact that we cannot clone qubits can be used to our advantage, preventing unwanted copies from being made.

The second comment is so important it deserves its own section.

Quantum Computation versus Classical Computation

The qubits $|0\rangle$ and $|1\rangle$ correspond to the bits 0 and 1. If we run our quantum *CNOT* gate just using the qubits $|0\rangle$ and $|1\rangle$, and not any superpositions, then the computation is exactly the same as running a classical *CNOT* gate with 0 and 1. The same is true of the quantum version of the Fredkin gate. Since the classical Fredkin gate is universal and the quantum Fredkin gate using just $|0\rangle$ and $|1\rangle$ is equivalent to the classical gate, we can see that a quantum circuit can calculate anything that can be calculated by a classical circuit. The no-cloning property may seem worrisome, but it doesn't restrict us from doing classical computations in any way.

This is a deep result. It shows that if we compare classical and quantum computation, we shouldn't think of them as different types of computation. Quantum computation includes all of classical computation. It is the more general form of computation. The qubit is the basic unit of computation, not the bit.

Now that we have seen some basic gates, we will start to connect them together to form circuits.

The Bell Circuit

We call the following quantum circuit the Bell circuit.

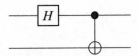

To see what it does, we will input the four pairs of qubits that form the standard basis. We start with $|00\rangle = |0\rangle|0\rangle$. The first qubit is acted on by the Hadamard gate's changing it to $\frac{1}{\sqrt{2}}|0\rangle + \frac{1}{\sqrt{2}}|1\rangle$, so the system of two qubits has state

$$\left(\frac{1}{\sqrt{2}}|0\rangle + \frac{1}{\sqrt{2}}|1\rangle\right)|0\rangle = \frac{1}{\sqrt{2}}|00\rangle + \frac{1}{\sqrt{2}}|10\rangle$$

at this stage. We now apply the $CNOT$ gate. This flips $|10\rangle$ to $|11\rangle$, giving the final state $\frac{1}{\sqrt{2}}|00\rangle + \frac{1}{\sqrt{2}}|11\rangle$.

We can represent the situation by the following picture.

$$|0\rangle \quad \boxed{H} \quad\bullet\quad$$
$$|0\rangle \quad\quad \oplus \quad \Big\}\ \tfrac{1}{\sqrt{2}}|00\rangle + \tfrac{1}{\sqrt{2}}|11\rangle$$

We will summarize this by

$$B(|00\rangle) = \frac{1}{\sqrt{2}}|00\rangle + \frac{1}{\sqrt{2}}|11\rangle.$$

Convince yourself that

$$B(|01\rangle) = \frac{1}{\sqrt{2}}|01\rangle + \frac{1}{\sqrt{2}}|10\rangle.$$

$$B(|10\rangle) = \frac{1}{\sqrt{2}}|00\rangle - \frac{1}{\sqrt{2}}|11\rangle.$$

$$B(|11\rangle) = \frac{1}{\sqrt{2}}|01\rangle - \frac{1}{\sqrt{2}}|10\rangle.$$

Each of these outputs is entangled. Since the inputs form an orthonormal basis for $\mathbb{R}^4$, the outputs must also form an orthonormal basis. This basis, consisting of four entangled kets, is called the Bell basis.

Recall that the way to tell whether a square matrix A is orthogonal is by calculating $A^T A$, where A^T is the transpose matrix obtained from A by interchanging the rows and columns. If we get the identity matrix I, then the matrix is orthogonal and the columns of the matrix give us an orthonormal basis. If we don't get the identity, then the matrix is not orthogonal. We have defined our gates to be orthogonal, so they all have this property. In fact, all the gates we have introduced in this chapter, with the one exception of the Pauli matrix Y, also have the property that when you take the transpose matrix you end up with exactly the same matrix you started with.** Consequently, for all of these gates, $AA = I$. This tells us that if we apply the gate twice in a row we end up with an output that is unchanged from the input. The second time we apply the gate, it undoes what we did when we applied it the first time.

We will see a couple of uses of the Bell circuit in a moment, but first we make use of the fact that the Hadamard gate and the $CNOT$ gate are their own inverses. Consider the following circuit:

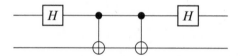

If we send a pair of qubits through the circuit, the first thing that happens is that the Hadamard gate is applied, and then we apply the $CNOT$ gate. This action is immediately undone by the second application of the $CNOT$ gate. Finally, the second application of the Hadamard gate undoes the action done by the initial Hadamard gate. The result is that the circuit doesn't change anything. The qubits output are identical to the qubits that entered. The second half of the circuit reverses what the first half does.

This means that the following circuit, which we will call the reverse Bell circuit, reverses the action of the Bell circuit.

** Matrices with the property that $A^T = A$ are called *symmetric*. They are symmetric with respect to the main diagonal.

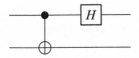

In particular, we know what happens if we input vectors from the Bell basis. It is going to give us vectors in the standard basis.

If we input $\frac{1}{\sqrt{2}}|00\rangle + \frac{1}{\sqrt{2}}|11\rangle$, it will output $|00\rangle$.

If we input $\frac{1}{\sqrt{2}}|01\rangle + \frac{1}{\sqrt{2}}|10\rangle$, it will output $|01\rangle$.

If we input $\frac{1}{\sqrt{2}}|00\rangle - \frac{1}{\sqrt{2}}|11\rangle$, it will output $|10\rangle$.

If we input $\frac{1}{\sqrt{2}}|01\rangle - \frac{1}{\sqrt{2}}|10\rangle$, it will output $|11\rangle$.

Now that we have the basic properties of the Bell circuit, we will see how it can be applied to do some very interesting things. We look at superdense coding and quantum teleportation.

Superdense Coding

The initial setup for both superdense coding and quantum teleportation is the same. Two electrons have the entangled spin state $\frac{1}{\sqrt{2}}|00\rangle + \frac{1}{\sqrt{2}}|11\rangle$. One of the electrons is given to Alice and the other to Bob. They then travel far apart, both being careful not to make any measurement of their respective electron, preserving the entangled state.

In superdense coding, Alice wants to send Bob two classical bits of information, that is, one out of the following possibilities: 00, 01, 10, 11. She is going to do this by sending Bob one qubit—her electron. We will outline the exact procedure in a moment, but first we will analyze the problem to see what we want to do.

Initially, it seems as though the solution should be easy. Alice is going to send Bob a qubit $a_0|0\rangle + a_1|1\rangle$. There are infinitely many choices for the qubit, anything that satisfies $a_0^2 + a_1^2 = 1$ will do. Surely, it must be easy to construct a way of transmitting two bits of information—one out of four possibilities—if you are allowed to send something that can be one of an infinite number of things. The problem is, of course, that Bob can never know what the qubit is. He can get information only by measuring. He will measure the spin in the standard basis and get either $|0\rangle$ or $|1\rangle$. If Alice sends him $a_0|0\rangle + a_1|1\rangle$, he will get $|0\rangle$ with probability a_0^2 and $|1\rangle$ with probability a_1^2. If he gets $|0\rangle$, he knows nothing about a_0, except for the fact that it is nonzero. Bob can get at most one bit of information from each qubit. In order to get two bits of information he will have to extract one bit from the particle that Alice is sending him, but he must also extract one bit from the particle in his possession.

Alice and Bob initially have one electron each. Eventually Bob is going to have both electrons and is going to measure their spins. Bob will have some quantum circuit with two wires exiting. If Alice wants to send 00, we need to arrange things so that just before Bob starts measuring, the top electron is in state $|0\rangle$ and the bottom electron is in state $|0\rangle$, that is, the pair of electrons is in the unentangled state $|00\rangle$ just before Bob measures their spins. Similarly, if Alice wants to send 01, we want the pair of electrons to be in the state $|01\rangle$ just before Bob makes his measurements. The final state should be $|10\rangle$ if Alice wants to send 10, and $|11\rangle$ if Alice wants to send 11.

The final observation is that Bob must do the same thing to every pair of electrons that he receives. He cannot do different things depending on what Alice is trying to send, because he doesn't know what she is trying to send. That's the whole point!

The idea behind the method is that Alice will act on her electron in one of four ways. Each way will result in the state of the qubits being one of the basis vectors in the Bell basis. Bob will then run the pair of qubits through the reverse Bell circuit to get the correct unentangled state.

Alice has four quantum circuits, one for each of the two-bit choices. Each circuit uses Pauli gates. The circuits are given below.

Circuit for 00   Circuit for 01

Circuit for 10   Circuit for 11

We will look at what happens to the qubits in each case. Initially, Alice's and Bob's qubits are entangled. They are in state $\frac{1}{\sqrt{2}}|00\rangle + \frac{1}{\sqrt{2}}|11\rangle$ which we will write as

$$\frac{1}{\sqrt{2}}|0\rangle \otimes |0\rangle + \frac{1}{\sqrt{2}}|1\rangle \otimes |1\rangle .$$

When Alice sends her electron through the appropriate circuit, her kets change. Note that Alice's circuits do not affect Bob's electron in any way. We will do the calculation in each case.

If Alice wants to send 00, then she does nothing. The resultant state for the qubits remains as state $\frac{1}{\sqrt{2}}|00\rangle + \frac{1}{\sqrt{2}}|11\rangle$.

If Alice wants to send 01, she applies X. This interchanges her $|0\rangle$ and her $|1\rangle$. The new state will be $\frac{1}{\sqrt{2}}|1\rangle \otimes |0\rangle + \frac{1}{\sqrt{2}}|0\rangle \otimes |1\rangle$, which we can write as $\frac{1}{\sqrt{2}}|10\rangle + \frac{1}{\sqrt{2}}|01\rangle$.

If Alice wants to send 10, she applies Z. This interchanges leaves $|0\rangle$ alone but changes her $|1\rangle$ to $-|1\rangle$. The new state will be $\frac{1}{\sqrt{2}}|0\rangle \otimes |0\rangle + \frac{1}{\sqrt{2}}(-|1\rangle) \otimes |1\rangle$ which we can write as $\frac{1}{\sqrt{2}}|00\rangle - \frac{1}{\sqrt{2}}|11\rangle$.

If Alice wants to send 11, she applies Y. The qubits end in the entangled state $\frac{1}{\sqrt{2}}|10\rangle - \frac{1}{\sqrt{2}}|01\rangle$.

Notice that these resultant states are exactly what she wants. Each is a distinct Bell basis vector. Now she sends Bob her electron. When Bob has her electron, he can use a circuit that inputs both the qubit that Alice has sent and the one that has always been in his possession. He uses the reverse Bell circuit.

If Alice is sending 00, when Bob receives the qubits they will be in state $\frac{1}{\sqrt{2}}|00\rangle + \frac{1}{\sqrt{2}}|11\rangle$. He sends this through the reverse Bell circuit. This changes the state to $|00\rangle$. This is unentangled. The top bit is $|0\rangle$ as is the bottom bit. Bob now measures the qubits. He gets 00.

If Alice is sending 01, when Bob receives the qubits they will be in state $\frac{1}{\sqrt{2}}|10\rangle + \frac{1}{\sqrt{2}}|01\rangle$ He sends this through the reverse Bell circuit. This changes the state to $|01\rangle$. This is unentangled. The top bit is $|0\rangle$ and the bottom bit is $|1\rangle$. Bob now measures the qubits. He gets 01. The other cases are similar. In each case Bob ends up with the two bits that Alice wants to send to him.

Quantum Teleportation

As in superdense coding, Alice and Bob are far apart. They each have one electron. The electrons share the entangled state $\frac{1}{\sqrt{2}}|00\rangle + \frac{1}{\sqrt{2}}|11\rangle$. Alice also has another electron. It is in state $a|0\rangle + b|1\rangle$. Alice has no idea what the probability amplitudes a and b are, but she and Bob want to change Bob's electron so that it has state $a|0\rangle + b|1\rangle$. They want to teleport the state of Alice's electron to Bob's. To do this, we will see that Alice needs to send Bob two classical bits, but notice that there are infinitely many possibilities for the initial state of her electron. It's impressive that we can send one of an infinite number of possibilities using only two classical bits. It is also interesting that Alice starts with a qubit and Bob ends up with it, but neither of them can ever know exactly what it is. To learn about it, they have to make a measurement. When they measure, they just get either $|0\rangle$ or $|1\rangle$.

We can deduce a few things about how the process will work. Bob is going to end up with an electron in the unentangled state $a|0\rangle + b|1\rangle$. At the start, Bob and Alice's electrons share an entangled state. To disentangle the state someone has to make a measurement. Clearly, it cannot be Bob. If Bob makes a measurement he will end up with an electron in state of either $|0\rangle$ or $|1\rangle$, not the required $a|0\rangle + b|1\rangle$, so we know Alice will be making a measurement. We also have to get the third electron's state involved. Alice will have to do something to entangle the state of this electron with the state of her other electron, which is currently entangled with Bob's. The obvious way of doing this is to send the two qubits that she controls through a *CNOT* gate. That will be the first step. The second step will be to apply the Hadamard gate to the top qubit. So, in fact, Alice is going to put the two qubits that she controls through a reverse Bell circuit. The situation is depicted as follows, where Alice's qubits are shown above Bob's qubit. The second and third rows depict the entangled qubits.

$a|0\rangle + b|1\rangle$ — Alice's qubits

$\frac{1}{\sqrt{2}}|00\rangle + \frac{1}{\sqrt{2}}|11\rangle$ — Bob's qubit

We have three qubits, the initial state that describes the three electrons is

$$(a|0\rangle + b|1\rangle) \otimes \left(\frac{1}{\sqrt{2}}|00\rangle + \frac{1}{\sqrt{2}}|11\rangle\right),$$

which we can write as

$$\frac{a}{\sqrt{2}}|000\rangle + \frac{a}{\sqrt{2}}|011\rangle + \frac{b}{\sqrt{2}}|100\rangle + \frac{b}{\sqrt{2}}|111\rangle.$$

Alice is going to act on her qubits, so we write the state emphasizing these.

$$\frac{a}{\sqrt{2}}|00\rangle \otimes |0\rangle + \frac{a}{\sqrt{2}}|01\rangle \otimes |1\rangle + \frac{b}{\sqrt{2}}|10\rangle \otimes |0\rangle + \frac{b}{\sqrt{2}}|11\rangle \otimes |1\rangle.$$

Alice is going to apply the reverse Bell circuit. We will analyze this in two steps, first by applying the *CNOT* gate to the first two qubits and then the Hadamard gate to the top bit. Applying the CNOT gate gives:

$$\frac{a}{\sqrt{2}}|00\rangle \otimes |0\rangle + \frac{a}{\sqrt{2}}|01\rangle \otimes |1\rangle + \frac{b}{\sqrt{2}}|11\rangle \otimes |0\rangle + \frac{b}{\sqrt{2}}|10\rangle \otimes |1\rangle.$$

Alice now is going to act on the first qubit, so we write the state emphasizing this.

$$\frac{a}{\sqrt{2}}|0\rangle \otimes |0\rangle \otimes |0\rangle + \frac{a}{\sqrt{2}}|0\rangle \otimes |1\rangle \otimes |1\rangle + \frac{b}{\sqrt{2}}|1\rangle \otimes |1\rangle \otimes |0\rangle + \frac{b}{\sqrt{2}}|1\rangle \otimes |0\rangle \otimes |1\rangle.$$

We now apply the Hadamard gate to the first qubit. This changes $|0\rangle$ to

$$\frac{1}{\sqrt{2}}|0\rangle + \frac{1}{\sqrt{2}}|1\rangle \quad \text{and } |1\rangle \text{ to } \quad \frac{1}{\sqrt{2}}|0\rangle - \frac{1}{\sqrt{2}}|1\rangle.$$

This results in the state

$$\frac{a}{2}|0\rangle \otimes |0\rangle \otimes |0\rangle + \frac{a}{2}|1\rangle \otimes |0\rangle \otimes |0\rangle + \frac{a}{2}|0\rangle \otimes |1\rangle \otimes |1\rangle$$

$$+ \frac{a}{2}|1\rangle \otimes |1\rangle \otimes |1\rangle + \frac{b}{2}|0\rangle \otimes |1\rangle \otimes |0\rangle - \frac{b}{2}|1\rangle \otimes |1\rangle \otimes |0\rangle$$

$$+ \frac{b}{2}|0\rangle \otimes |0\rangle \otimes |1\rangle - \frac{b}{2}|1\rangle \otimes |0\rangle \otimes |1\rangle$$

This can be slightly simplified to give

$$\frac{1}{2}|00\rangle \otimes (a|0\rangle + b|1\rangle) + \frac{1}{2}|01\rangle \otimes (a|1\rangle + b|0\rangle)$$
$$+ \frac{1}{2}|10\rangle \otimes (a|0\rangle - b|1\rangle) + \frac{1}{2}|11\rangle \otimes (a|1\rangle - b|0\rangle).$$

Alice now measures her two electrons in the standard basis. She will get one of $|00\rangle, |01\rangle, |10\rangle, |11\rangle$, each with probability 1/4.

If she gets $|00\rangle$, Bob's qubit will jump to state $a|0\rangle + b|1\rangle$.

If she gets $|01\rangle$, Bob's qubit will jump to state $a|1\rangle + b|0\rangle$.

If she gets $|10\rangle$, Bob's qubit will jump to state $a|0\rangle - b|1\rangle$.

If she gets $|11\rangle$, Bob's qubit will jump to state $a|1\rangle - b|0\rangle$.

Alice and Bob want Bob's qubit to be in the state $a|0\rangle + b|1\rangle$. It is almost there, but not quite. To sort things out, Alice has to let Bob know which of the four possible situations he is in. She sends Bob two classical bits of information, 00, 01, 10, or 11, corresponding to the results of her measurements, to let him know. These bits of information can be sent in any way, by text, for example.

If Bob receives 00, he knows that his qubit is in the correct form and so does nothing.

If Bob receives 01, he knows that his qubit is $a|1\rangle + b|0\rangle$. He applies the gate X to it.

If Bob receives 10, he knows that his qubit is $a|0\rangle - b|1\rangle$. He applies the gate Z to it.

If Bob receives 11, he knows that his qubit is $a|1\rangle - b|0\rangle$. He applies the gate Y to it.

In every case Bob's qubit ends in state $a|0\rangle + b|1\rangle$, the original state of the qubit that Alice wanted to teleport.

It is important to note that there is only one qubit in state $a|0\rangle + b|1\rangle$ at any point during the process. Initially, Alice has it. At the end Bob has it, but as the no cloning theorem tells us, we can't copy, so only one of them can have it at a time.

It is also interesting to observe that when Alice sends her qubits through her circuit Bob's qubit instantaneously jumps to one of the four states. He has to wait for Alice to send him the two classical bits before he can determine which of the four qubits correspond to Alice's original qubit. It is the fact that the two bits have to be sent by some conventional transportation method that prevents instantaneous transmission of information.

Quantum teleportation and superdense coding are sometimes described as being inverse operations. For superdense coding, Alice sends Bob one qubit to convey two classical bits of information. For quantum teleportation, Alice sends Bob two classical bits of information to teleport one qubit. For superdense coding, Alice encodes using the Pauli transformations, and Bob decodes using the reverse Bell circuit. For quantum teleportation, Alice encodes using the reverse Bell circuit, and Bob decodes using the Pauli transformations.

Quantum teleportation is actually being performed, usually using entangled photons rather than entangled electrons, where it can be done over substantial distances. As I write this, it has been announced that a Chinese team has teleported a qubit from Earth to a satellite in low Earth orbit. These experiments are often mentioned on news broadcasts, mainly because of the word "teleportation," which conjures up images of *Star Trek*. Unfortunately, quantum teleportation is not something that is readily explained in a brief sound bite, and though many people have heard the term, not many understand exactly what it is that is being teleported.

Quantum teleportation gives a way of transporting a qubit from one place to another without actually transporting the particle that represents the qubit. It is used in various ways to correct errors. This is extremely important for quantum computations. Qubits have a tendency to interact with the environment and get corrupted. We will not study error correction in detail but will only look at a simple example.

Error Correction

I was a student before the advent of CDs. We listened to vinyl records. To play a record we went through an elaborate ritual. First, the record was gently slid from its sleeve, care being taken to hold it by its edges and not get any fingerprints on the surface. Then the record was placed on the turntable. The next step was to clean it of any dust. This often involved an antistatic spray and a special cleaning brush. Finally you lined up the stylus and carefully lowered it to the record.

Even with all these precautions, there were often clicks and pops caused by unseen dust or some minute imperfection. If you accidently scratched it, you would get a pop thirty three times per minute, which made the music unlistenable. Then CDs came. Gone were the pops. You could even scratch the surface and it still played perfectly. It seemed incredible.

Vinyl records have no error correction built in. If you damage them, you cannot recover the original sound. CDs, on the other hand, incorporate error correction. If there is some small imperfection, the digital error-correcting code can often calculate what has gone wrong and correct it.

Encoding digital information involves two essential ideas. The first is to eliminate redundancy to compress the information as much as possible—to make the message as short as possible. A good example of this is making a ZIP file of a document. (Some people don't like CDs because they think the music has been compressed too much, losing the warmth you get from vinyl.) The second important idea is to add some redundancy back in, but to make it useful redundancy. You want to add in some additional information that will help correct errors.

Nowadays, practically all transmissions of digital information use some form of error-correcting code. There are so many ways that a message can be slightly corrupted, and given a slightly corrupted message, you want to be able to correct it.

Error correction is essential for transmissions involving qubits. We are using photons and electrons to encode them. These particles can interact with the rest of the universe and unwanted interactions may change the states of some qubits.

In this section we will look at the most basic classical error-correcting code and then show how it can be modified for sending qubits.

The Repetition Code

A simple error-correcting code is just to repeat the symbol that we want to send. The simplest case is to repeat it three times. If Alice wants to send 0, she sends 000. If she wants to send 1, she sends 111. If Bob keeps getting sequences of three 0s and three 1s, he assumes that all is well. If he receives something else, say 101, he knows that an error has occurred; the string should have been 000 or 111. If the string that Alice sent was 000, then two errors must have occurred. If the string was 111, then only one error has occurred. If errors are fairly unlikely, it is more probable that one error, rather than two errors, have occurred, so Bob assumes that the least number of errors have occurred and consequently replaces 101 with 111.

There are eight possibilities of three-bit strings that Bob could receive. Four of them are 000, 001, 010, and 100. Bob decodes all of these as 000. The other four three-bit strings are 111, 110, 101, and 011. Bob decodes these as

111. If the chance of error is very small, then this repetition code corrects many errors and reduces the overall error rate. This is fairly straightforward, but we will analyze what Bob does in a way that generalizes for qubits. The problem with qubits is that to read them, we have to measure them, and that can make them jump to a new state. We need a new way of determining what Bob should do. He is going to perform parity tests.

Now, suppose Bob receives the three bits $b_0 b_1 b_2$. We will do some computations to show which, if any, of the bits should be changed. Bob computes $b_0 \oplus b_1$ and $b_0 \oplus b_2$.

The first sum checks the parity of the first two bits—that is, it checks whether they are the same digit or not. The second sum performs a parity check on the first and third digits.

If all three bits equal 0, or all equal 1, then he will get 0 for both sums. If not all of the bits are equal, then two will be equal and the third will differ. It will be this third symbol that needs to be flipped from 0 to 1, or from 1 to 0.

If $b_0 = b_1 \neq b_2$, then $b_0 \oplus b_1 = 0$ and $b_0 \oplus b_2 = 1$.

If $b_0 = b_2 \neq b_1$, then $b_0 \oplus b_1 = 1$ and $b_0 \oplus b_2 = 0$.

If $b_0 \neq b_1 = b_2$, then $b_0 \oplus b_1 = 1$ and $b_0 \oplus b_2 = 1$.

This means that Bob can look at the pair of bits $b_0 \oplus b_1$ and $b_0 \oplus b_2$.

If he gets 00 then there is nothing to correct, so he does nothing.

If he gets 01, he flips b_2.

If he gets 10, he flips b_1.

If he gets 11, he flips b_0.

We look at how these error-correcting ideas can be modified for qubits. But before we do we make one important observation. It might seem trivial, but it is what makes the quantum bit-flip correction code work.

Suppose Bob receives a string and there is an error in the first bit. This means that he has received either 011 or 100. After Bob does the parity tests, he will get 11 for both strings and will know that there is an error in the first bit. The key observation is that the parity tests tell us where the error is. They do not tell us whether it is a 0 that needs to be flipped to a 1, or a 1 that needs to be flipped to a 0.

Quantum Bit-Flip Correction

Alice wants to send the qubit $a|0\rangle + b|1\rangle$ to Bob. There are various types of errors that can occur, but we will restrict our attention to bits getting flipped. In this case, $a|0\rangle + b|1\rangle$ gets changed to $a|1\rangle + b|0\rangle$.

Alice would like to send three copies of her qubit. This, of course, is not possible. The no cloning theorem tells us that she cannot make copies. But she can perform what is essentially a classical fan-out and replace $|0\rangle$ with $|000\rangle$ and $|1\rangle$ with $|111\rangle$. This is done with two *CNOT* gates. This is shown in the circuit below.

She starts with three qubits, the one she wants to encode and two ancilla bits that are both $|0\rangle$, so the initial state is $(a|0\rangle + b|1\rangle)|0\rangle|0\rangle = a|0\rangle|0\rangle|0\rangle + b|1\rangle|0\rangle|0\rangle$. The first *CNOT* gate changes it to $a|0\rangle|0\rangle|0\rangle + b|1\rangle|1\rangle|0\rangle$. The second gives us the required state $a|0\rangle|0\rangle|0\rangle + b|1\rangle|1\rangle|1\rangle$.

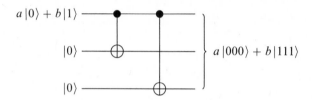

Alice then sends the three qubits to Bob. But the channel is noisy, and there is the possibility of a qubit being flipped. Bob might receive the correct qubits $a|000\rangle + b|111\rangle$, or he might receive one of the following incorrect versions, $a|100\rangle + b|011\rangle$, $a|010\rangle + b|101\rangle$ or $a|001\rangle + b|110\rangle$, which correspond to the error occurring in the first, second, and third qubit, respectively. He wants both to detect the error and to correct it. But notice that he cannot make any measurements on this entangled state. If he does, the state immediately becomes unentangled and he just gets three qubits that are some combination of $|0\rangle$s and $|1\rangle$s—the values of a and b are lost, with no way of recovering them.

It is amazing that Bob can determine which bit is flipped, correct it, and yet never make a measurement on the three qubits that Alice sent him! But he can. He uses the parity check idea that we used for classical bits.

He adds an additional two qubits in which to perform the parity checks. The circuit is given below. It uses four *CNOT* gates. The two on the fourth wire are used to do the $b_0 \oplus b_1$ parity calculation; the two on the fifth wire do the $b_0 \oplus b_2$ calculation. The standard first reaction on seeing this circuit is to assume that we end up with five qubits that are hopelessly entangled. But I've drawn the picture that shows that the bottom two qubits are not entangled with the top three. Can that really be the case?

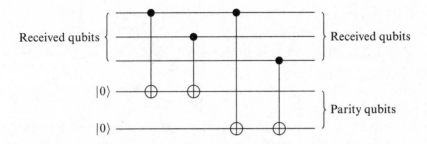

Let us suppose that Bob receives $a|c_0c_1c_2\rangle + b|d_0d_1d_2\rangle$ The key observation is that if there is an error, then there will be an error in both $c_0c_1c_2$ and $d_0d_1d_2$, and it will occur in exactly the same place. When we apply the parity checks, both strings give the same results.

To illustrate what is going on, let's look at Bob's circuit, ignoring the fifth wire for the moment. The input for the first four qubits is

$$(a|c_0c_1c_2\rangle + b|d_0d_1d_2\rangle)|0\rangle = a|c_0c_1c_2\rangle|0\rangle + b|d_0d_1d_2\rangle|0\rangle.$$

The two *CNOT* gates attached to the fourth wire perform the parity check on the first two digits. But $c_0 \oplus c_1 = d_0 \oplus d_1$, so the four qubits at the right of the circuit will be in one of two states. They will be in state

$$a|c_0c_1c_2\rangle|0\rangle + b|d_0d_1d_2\rangle|0\rangle = (a|c_0c_1c_2\rangle + b|d_0d_1d_2\rangle)|0\rangle$$

if $c_0 \oplus c_1 = d_0 \oplus d_1 = 0$.

They will be in state

$$a|c_0c_1c_2\rangle|1\rangle + b|d_0d_1d_2\rangle|1\rangle = (a|c_0c_1c_2\rangle + b|d_0d_1d_2\rangle)|1\rangle$$

if $c_0 \oplus c_1 = d_0 \oplus d_1 = 1$.

In both cases, the fourth qubit is not entangled with the top three.

A similar argument applies to the fifth qubit. It is not entangled with the others. It is $|0\rangle$ if $c_0 \oplus c_2 = d_0 \oplus d_2 = 0$, and is $|1\rangle$ if $c_0 \oplus c_2 = d_0 \oplus d_2 = 1$.

Since the bottom two qubits are not entangled with the top three, Bob can make measurements on the bottom two qubits, and it will leave the top three unchanged. This is what he does:

If he gets 00, then there is nothing to correct, so he does nothing.

If he gets 01, he flips the third qubit by installing an *X* gate on the third wire.

If he gets 10, he flips the second qubit using by installing an X gate on the second wire.

If he gets 11, he flips the first qubit using by installing an X gate on the first wire.

The result is that the bit-flip error is corrected and the qubits are now back in the state that Alice sent.

In this chapter we introduced the idea of quantum gates and circuits. We have seen some surprising things we can do with just a few quantum gates. We have also seen that quantum computation includes all of classical computation. This doesn't mean that we will be using quantum computers to perform classical computations, but it does tell us that quantum computation is the more fundamental form of computation.

The next topic that we look at concerns whether we can use quantum circuits to perform calculations faster than can be done with classical circuits. How do we measure the speed of a computation? Are quantum computers always faster than classical ones? These are some of the questions we look at in the next chapter.

8   Quantum Algorithms

Popular descriptions of quantum algorithms describe them as being much faster than regular algorithms. This speedup, it is explained, comes from being able to put the input into a superposition of all possible inputs and then performing the algorithm on the superposition. Consequently, instead of running the algorithm on just one input, as you do classically, you can run the algorithm using "quantum parallelism" on all possible inputs at the same time. These descriptions often end at this point. But this leaves many unanswered questions. We seem to end up with many possible answers all superimposed on one another. If we make a measurement, won't we get just one of these answers at random? There are far more likely to be wrong answers than right answers, so aren't we more likely to end up with a wrong answer than with the right answer?

Clearly, there has to be more to quantum algorithms than just putting everything into a superposition of states. The real art of constructing these algorithms is being able to manipulate these superpositions so that when we make measurements we get a useful answer. In this chapter, we will look at three quantum algorithms and see how they tackle this problem. We will see that not every algorithm is susceptible to a quantum speedup. Quantum algorithms are not classical algorithms that have been sped up. Instead, they involve quantum ideas to see the problem in a new light; the algorithms work not by the use of brute force, but by ingenious ways of exploiting underlying patterns that can be seen from only the quantum viewpoint.

We will describe three algorithms in detail. All three are ingenious exploitations of underlying mathematical patterns. The level of difficulty increases as we move through the algorithms. Some mathematics books use a star to denote a difficult section and a double star to denote a very difficult section. The Deutsch-Jozsa algorithm probably deserves a star, and Simon's algorithm a double star.

At the end of the chapter, we will talk a little about the properties that questions must have in order for a quantum algorithm to solve them faster than a classical one, and why they seem so hard! But first we must describe how the speed of algorithms is measured.

The Complexity Classes *P* and *NP*

Imagine that you are given the following problems. You are told that you are not allowed to use a calculator or computer but have to work them out using paper and pencil.

- Find two whole numbers bigger than 1 whose product is equal to 35.
- Find two whole numbers bigger than 1 whose product is equal to 187.
- Find two whole numbers bigger than 1 whose product is equal to 2,407.
- Find two whole numbers bigger than 1 whose product is equal to 88,631.

You won't have much difficulty doing the first question, but each subsequent question is harder and will take more steps and consequently more time to solve. Before we analyze this in more detail, consider another four problems.

- Multiply 7 by 5 and check that it equals 35.
- Multiply 11 by 17 and check that it equals 187.
- Multiply 29 by 83 and check that it equals 2407.
- Multiply 337 by 263 and check that it equals 88,631.

These questions are undoubtedly easier than the first series. Again each subsequent question takes more time to solve than the previous one, but the amount of time is growing more slowly. Even the fourth question takes less than a minute to solve by hand.

We will denote the number of digits of the input number by n, so in the first set of questions we start with $n = 2$ and go up to $n = 5$.

We will let $T(n)$ denote the time, or equivalently the number of steps, to solve a question of input length n. Complexity looks at how the size of $T(n)$ grows as n grows. In particular, we ask if we can find some positive numbers k and p such that $T(n) \leq kn^p$ for every value of n. If we can, we say that the underlying problem can be solved in *polynomial time*. If, on the other hand, we can find positive number k and a number $c > 1$, such that $T(n) > kc^n$ for every value of n, we say that the problem requires *exponential time*. Recall the basic fact concerning polynomial versus exponential growth: Given enough time, something with exponential growth will grow much faster than something with polynomial growth. In computer science, questions that can be solved in polynomial time are considered tractable, but those with exponential growth are not. Problems that can be solved in polynomial time are regarded as easy; those that require exponential time are hard. In practice, it turns out that most polynomial time problems involve polynomials with small degree, so even if we don't have the computational power to solve a problem with a large value of n at the moment, we should have it in a few years. On the other hand, with an exponential time problem, once the size has increased beyond what we can currently tackle, increasing the size of n even slightly more produces a problem that becomes much harder and is unlikely to be solvable in the foreseeable future.

Let's look at our two sets of problems. The second set involves multiplying two numbers together, but this is easy to do. As n increases it does take more time, but it can be shown that this is a polynomial time problem. What about the first set of questions? If you tried tackling them, you will probably believe that the amount of time needed is exponential in n and not polynomial in n, but is this the case? Everybody thinks so, but, on the other hand, nobody has found a proof.

In 1991, RSA Laboratories posted a challenge. It listed large numbers, each of which was the product of two primes. The challenge was to factor them. They went from being about 100 decimal digits long to 600 digits. You were of course allowed to use computers! There were prizes for the first person to factor them. The 100 digit numbers were factored relatively quickly, but the numbers with 300 or more digits still haven't been factored.

If a problem can be solved in polynomial time we say it belongs to the complexity class P. So the problem that consists of multiplying two numbers together belongs to P. Suppose that instead of solving the problem, someone gives you the answer and you just have to check that the answer is correct. If this process of checking that an answer is correct takes polynomial time, then we say the problem belongs to complexity class NP.* The

---

* NP comes from *nondeterministic polynomial*, which in turn refers to certain types of Turing machines that are called nondeterministic Turing machines.

problem of factoring a large number into the product of two primes belongs to *NP*.

Clearly, checking that an answer is correct is easier than actually finding the answer, so every problem that is in *P* is also in *NP*, but what about the converse question. Does every *NP* problem belong to *P*? Is it true that every question whose answer can be checked in polynomial time can also be solved in polynomial time? You are probably saying to yourself, "Of course not!" Most people would agree that it seems extremely unlikely, but nobody has managed to prove that *P* is not equal to *NP*. The problem of factoring a large number into the product of two primes belongs to *NP*, and we don't think it belongs to *P*, but nobody has been able to prove it.

The problem of whether *NP* is equal to *P* is one of the most important in computer science. In 2000, the Clay Mathematics Institute listed seven "Millennium Prize Problems," each with a prize of a million dollars. The *P* versus *NP* problem is one of the seven.

Are Quantum Algorithms Faster Than Classical Ones?

Most quantum computer scientists believe that *P* is not equal to *NP*. They also think that there are problems that are in *NP* but not *P*, which a quantum computer can solve in polynomial time. This means that there are problems that a quantum computer can solve in polynomial time that a classical computer cannot. To prove this, however, involves the first step of showing that some problem belongs to *NP* but not to *P*, and as we have seen, nobody knows how to do this. So, how can we compare the speed of quantum algorithms to classical algorithms? There are two ways: one theoretical, the other practical. The theoretical way is to invent a new way of measuring complexity that makes it easier to construct proofs. The practical way is to construct quantum algorithms for solving important real-world problems in polynomial time that we believe, but have been unable to prove, do not belong to *P*.

An example of the second approach is Shor's algorithm for factoring the product of two primes. Peter Shor constructed a quantum algorithm that works in polynomial time. We believe, but have been unable to prove, that a classical algorithm cannot do this in polynomial time. Why is that important? Well, as we shall see our Internet security depends on this. That said, in the rest of this chapter we will take the first approach—defining a new way of calculating complexity.

Query Complexity

All of the algorithms that we are going to look at in this chapter concern evaluating functions. The Deutsch and Deutsch-Jozsa algorithms consider functions that belong to two classes. We are given a function at random, and we have to determine which of the two classes the function belongs to. Simon's algorithm concerns periodic functions of a special type. Again we are given one of these functions at random, and we have to determine the period.

When we run these algorithms we have to evaluate the functions. The *query complexity* counts the number of times that we have to evaluate the function to get our answer. The function is sometimes called a *black box* or an *oracle*. Instead of saying that we are evaluating the function, we say that we are querying the black box or the oracle. The point of this is that we don't have to worry about how to write an algorithm that emulates the function, so we don't have to calculate the number of steps that function takes to evaluate the input. We just keep track of the number of questions. This is much simpler. To illustrate, we begin with the most elementary example.

Deutsch's Algorithm

David Deutsch is one of the founders of quantum computing. In 1985, he published a landmark paper that described quantum Turing machines and quantum computation.** This paper also includes the following algorithm—the first to show that a quantum algorithm could be faster than a classical one.

The problem concerns functions of just one variable. The input can be either 0 or 1. The output also just takes the values of 0 or 1. There are four of these functions that we will denote f_0, f_1, f_2, and f_3:

The function f_0 sends both inputs to 0; i.e., $f_0(0) = 0$ and $f_0(1) = 0$.

The function f_1 sends 0 to 0 and 1 to 1; i.e., $f_1(0) = 0$ and $f_1(1) = 1$.

** "Quantum theory, the Church-Turing principle and the universal quantum computer," *Proceedings of the Royal Society A* 400 (1818): 97–117.

The function f_2 sends 0 to 1 and 1 to 0; i.e., $f_2(0) = 1$ and $f_2(1) = 0$.

The function f_3 sends both inputs to 1; i.e., $f_3(0) = 1$ and $f_3(1) = 1$.

The functions f_0 and f_3 are called *constant* functions. The output is the same value for both inputs—the output is constant. A function is called *balanced* if it sends half its inputs to 0 and the other half to 1. Both f_1 and f_2 are balanced.

The question that Deutsch posed is this: Given one of these four functions at random, how many function evaluations do we have to make to determine whether the function is constant or balanced? It is important to understand what we are asking. We are not interested in which of the four functions we have, but solely in whether the given function is constant or not.

The classical analysis is as follows. We can evaluate our given function at either 0 or 1. Supposing that we choose to evaluate it by plugging in 0, then there are two possible outcomes—either we get 0 or we get 1. If we get 0, all we know is $f(0) = 0$. The function could be either f_0 or f_1. Since one is constant and the other is balanced, we are forced to evaluate our function again to decide between them. Classically, to answer the question we have to plug both 0 and 1 into the function. We need to make two function evaluations.

We now look at the quantum version of the question. First, we construct gates that correspond to the four functions. The following picture depicts the gates, where i can take on the numbers 0, 1, 2, or 3.

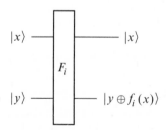

This says that:

If we input $|0\rangle \otimes |0\rangle$, it outputs $|0\rangle \otimes |f_i(0)\rangle$.

If we input $|0\rangle \otimes |1\rangle$, it outputs $|0\rangle \otimes |f_i(0) \oplus 1\rangle$.

If we input $|1\rangle \otimes |0\rangle$, it outputs $|1\rangle \otimes |f_i(1)\rangle$.

If we input $|1\rangle \otimes |1\rangle$, it outputs $|1\rangle \otimes |f_i(1) \oplus 1\rangle$.

Notice that for each i, one of $f_i(0)$ and $f_i(0) \oplus 1$ is equal to 0 and the other is equal to 1, and one of $f_i(1)$ and $f_i(1) \oplus 1$ is equal to 0 and the other is equal to 1. This means that the four outputs always give us the standard basis elements, telling us the matrix representing our gate is orthogonal—and so we really do have a gate.

Though we enter two bits of information and get two bits as output, the information these gates give for classical bits, $|0\rangle$ and $|1\rangle$ is exactly the same as for the functions evaluated at 0 and 1. The top qubit is exactly what we entered, so that piece of output gives us no new information. The choice of $|0\rangle$ and $|1\rangle$ for the second input gives us the option of the second output giving us the function evaluated on the top input ket or of the opposite answer. If we know one of these answers, we know the other.

The quantum computing question that corresponds to the classical question is: Given one of these four gates at random, how many times do you have to use the gate to determine whether the underlying function f_i is constant or whether it is balanced?

If we restrict to just entering $|0\rangle$ and $|1\rangle$ into the gate, the analysis is exactly the same as before. You have to use the gate twice. But David Deutsch showed that if we are allowed to input qubits containing superpositions of $|0\rangle$ and $|1\rangle$, the gate only needs to be used once. To show this, he used the following circuit.

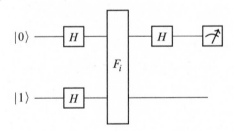

The little meter symbol at the right end of the top wire means that we are going to measure this qubit. The lack of the meter symbol on the second wire tells us that we won't be measuring the second output qubit. Let us see how this circuit works.

The qubits $|0\rangle \otimes |1\rangle$ are input. They go through the Hadamard gates, which puts them in the state

$$\frac{1}{\sqrt{2}}(|0\rangle + |1\rangle) \otimes \frac{1}{\sqrt{2}}(|0\rangle - |1\rangle) = \frac{1}{2}(|00\rangle - |01\rangle + |10\rangle - |11\rangle).$$

These then go through the F_i gate. The state becomes

$$\frac{1}{2}\left(|0\rangle \otimes |f_i(0)\rangle - |0\rangle \otimes |f_i(0) \oplus 1\rangle + |1\rangle \otimes |f_i(1)\rangle - |1\rangle \otimes |f_i(1) \oplus 1\rangle\right).$$

This can be rearranged to give:

$$\frac{1}{2}\left(|0\rangle \otimes \left(|f_i(0)\rangle - |f_i(0) \oplus 1\rangle\right) + |1\rangle \otimes \left(|f_i(1)\rangle - |f_i(1) \oplus 1\rangle\right)\right).$$

We now make the observation that $|f_i(0)\rangle - |f_i(0) \oplus 1\rangle$ is either $|0\rangle - |1\rangle$ or $|1\rangle - |0\rangle$, depending on whether $f_i(0)$ is 0 or 1. But we can be clever about this and write

$$|f_i(0)\rangle - |f_i(0) \oplus 1\rangle = (-1)^{f_i(0)}\left(|0\rangle - |1\rangle\right).$$

We also deduce that

$$|f_i(1)\rangle - |f_i(1) \oplus 1\rangle = (-1)^{f_i(1)}\left(|0\rangle - |1\rangle\right).$$

The state of our qubits after passing through the F_i gate can then be written as

$$\frac{1}{2}\left(|0\rangle \otimes \left((-1)^{f_i(0)}\left(|0\rangle - |1\rangle\right)\right) + |1\rangle \otimes \left((-1)^{f_i(1)}\left(|0\rangle - |1\rangle\right)\right)\right).$$

We can rearrange this to give

$$\frac{1}{2}\left((-1)^{f_i(0)}|0\rangle \otimes \left((|0\rangle - |1\rangle)\right) + (-1)^{f_i(1)}|1\rangle \otimes \left((|0\rangle - |1\rangle)\right)\right),$$

then

$$\frac{1}{2}\left((-1)^{f_i(0)}|0\rangle + (-1)^{f_i(1)}|1\rangle\right) \otimes \left(|0\rangle - |1\rangle\right),$$

and finally

$$\frac{1}{\sqrt{2}}\left((-1)^{f_i(0)}|0\rangle + (-1)^{f_i(1)}|1\rangle\right) \otimes \frac{1}{\sqrt{2}}\left(|0\rangle - |1\rangle\right).$$

This shows that the two qubits are not entangled, and the top qubit has state

$$\frac{1}{\sqrt{2}}\left((-1)^{f_i(0)}|0\rangle + (-1)^{f_i(1)}|1\rangle\right).$$

Let's examine this state for each of the four possibilities for f_i.

For f_0, we have $f_0(0) = f_0(1) = 0$, so the qubit is $(1/\sqrt{2})(|0\rangle + |1\rangle)$.

For f_1, we have $f_1(0) = 0$ and $f_1(1) = 1$, so the qubit is $(1/\sqrt{2})(|0\rangle - |1\rangle)$.

For f_2, we have $f_2(0) = 1$ and $f_2(1) = 0$, so the qubit is $(-1/\sqrt{2})(|0\rangle - |1\rangle)$.

For f_3, we have $f_3(0) = f_3(1) = 1$, so the qubit is $(-1/\sqrt{2})(|0\rangle + |1\rangle)$.

The next step in the circuit is to send our qubit through the Hadamard gate. This gate sends $(1/\sqrt{2})(|0\rangle + |1\rangle)$ to $|0\rangle$ and $(1/\sqrt{2})(|0\rangle - |1\rangle)$ to $|1\rangle$. So we know:

If $i = 0$, the qubit is $|0\rangle$.

If $i = 1$, the qubit is $|1\rangle$.

If $i = 2$, the qubit is $-|1\rangle$.

If $i = 3$, the qubit is $-|0\rangle$.

If we now measure the qubit in the standard basis, we will get 0 if i is either 0 or 3, and we will get 1 if i is either 1 or 2. Of course, f_0 and f_3 are the constant functions and f_1 and f_2 are the balanced. So, if after measuring we get 0, we know with certainty that the original function was constant. If we get 1, we know that the original function was balanced.

Consequently, we need to ask the oracle only one question versus two. For Deutsch's problem there is therefore a slight speedup using a quantum algorithm. This algorithm has no real practical applications, but, as we noted earlier, it was the first example of proving that there are quantum algorithms faster than classical ones.

We will look at two other quantum algorithms in detail. They both involve inputting a number of qubits and then sending each one through a Hadamard gate. We introduce a little more mathematics to help keep the description of many qubits in superposition from becoming too unwieldy.

The Kronecker Product of Hadamard Matrices

We know that the matrix for the Hadamard gate is given by

$$H = \begin{bmatrix} \dfrac{1}{\sqrt{2}} & \dfrac{1}{\sqrt{2}} \\ \dfrac{1}{\sqrt{2}} & -\dfrac{1}{\sqrt{2}} \end{bmatrix} = \frac{1}{\sqrt{2}}\begin{bmatrix} 1 & 1 \\ 1 & -1 \end{bmatrix}.$$

This tells us that

$$H(|0\rangle) = \frac{1}{\sqrt{2}}\begin{bmatrix} 1 & 1 \\ 1 & -1 \end{bmatrix}\begin{bmatrix} 1 \\ 0 \end{bmatrix} = \frac{1}{\sqrt{2}}\begin{bmatrix} 1 \\ 1 \end{bmatrix} = \frac{1}{\sqrt{2}}\begin{bmatrix} 1 \\ 0 \end{bmatrix} + \frac{1}{\sqrt{2}}\begin{bmatrix} 0 \\ 1 \end{bmatrix} = \frac{1}{\sqrt{2}}|0\rangle + \frac{1}{\sqrt{2}}|1\rangle,$$

and

$$H(|1\rangle) = \frac{1}{\sqrt{2}}\begin{bmatrix} 1 & 1 \\ 1 & -1 \end{bmatrix}\begin{bmatrix} 0 \\ 1 \end{bmatrix} = \frac{1}{\sqrt{2}}\begin{bmatrix} 1 \\ -1 \end{bmatrix} = \frac{1}{\sqrt{2}}\begin{bmatrix} 1 \\ 0 \end{bmatrix} - \frac{1}{\sqrt{2}}\begin{bmatrix} 0 \\ 1 \end{bmatrix} = \frac{1}{\sqrt{2}}|0\rangle - \frac{1}{\sqrt{2}}|1\rangle \ .$$

Suppose that we input two qubits and send both through Hadamard gates. The four basis vectors will be sent as follows:

$|0\rangle \otimes |0\rangle$ goes to

$$\left(\frac{1}{\sqrt{2}}|0\rangle + \frac{1}{\sqrt{2}}|1\rangle \right) \otimes \left(\frac{1}{\sqrt{2}}|0\rangle + \frac{1}{\sqrt{2}}|1\rangle \right) = \frac{1}{2}(|00\rangle + |01\rangle + |10\rangle + |11\rangle).$$

$|0\rangle \otimes |1\rangle$ goes to

$$\left(\frac{1}{\sqrt{2}}|0\rangle + \frac{1}{\sqrt{2}}|1\rangle \right) \otimes \left(\frac{1}{\sqrt{2}}|0\rangle - \frac{1}{\sqrt{2}}|1\rangle \right) = \frac{1}{2}(|00\rangle - |01\rangle + |10\rangle - |11\rangle).$$

$|1\rangle \otimes |0\rangle$ goes to

$$\left(\frac{1}{\sqrt{2}}|0\rangle - \frac{1}{\sqrt{2}}|1\rangle \right) \otimes \left(\frac{1}{\sqrt{2}}|0\rangle + \frac{1}{\sqrt{2}}|1\rangle \right) = \frac{1}{2}(|00\rangle + |01\rangle - |10\rangle - |11\rangle).$$

$|1\rangle \otimes |1\rangle$ goes to

$$\left(\frac{1}{\sqrt{2}}|0\rangle - \frac{1}{\sqrt{2}}|1\rangle \right) \otimes \left(\frac{1}{\sqrt{2}}|0\rangle - \frac{1}{\sqrt{2}}|1\rangle \right) = \frac{1}{2}(|00\rangle - |01\rangle - |10\rangle + |11\rangle).$$

Recall that we can write everything in terms of four-dimensional kets. The previous four statements are equivalent to saying:

$$\begin{bmatrix} 1 \\ 0 \\ 0 \\ 0 \end{bmatrix} \text{ goes to } \frac{1}{2}\begin{bmatrix} 1 \\ 1 \\ 1 \\ 1 \end{bmatrix},$$

$$\begin{bmatrix} 0 \\ 1 \\ 0 \\ 0 \end{bmatrix} \text{ goes to } \frac{1}{2}\begin{bmatrix} 1 \\ -1 \\ 1 \\ -1 \end{bmatrix},$$

$$\begin{bmatrix} 0 \\ 0 \\ 1 \\ 0 \end{bmatrix} \text{ goes to } \frac{1}{2}\begin{bmatrix} 1 \\ 1 \\ -1 \\ -1 \end{bmatrix},$$

$$\begin{bmatrix} 0 \\ 0 \\ 0 \\ 1 \end{bmatrix} \text{ goes to } \frac{1}{2}\begin{bmatrix} 1 \\ -1 \\ -1 \\ 1 \end{bmatrix}.$$

This is a description of an orthonormal basis being sent to another orthonormal basis. So, we can write the matrix that corresponds to this. We call this new matrix $H^{\otimes 2}$.

$$H^{\otimes 2} = \frac{1}{2}\begin{bmatrix} 1 & 1 & 1 & 1 \\ 1 & -1 & 1 & -1 \\ 1 & 1 & -1 & -1 \\ 1 & -1 & -1 & 1 \end{bmatrix}$$

There is an underlying pattern to this matrix that involves H.

$$H^{\otimes 2} = \frac{1}{2}\begin{bmatrix} 1 & 1 & 1 & 1 \\ 1 & -1 & 1 & -1 \\ 1 & 1 & -1 & -1 \\ 1 & -1 & -1 & 1 \end{bmatrix} = \frac{1}{\sqrt{2}}\begin{bmatrix} \begin{bmatrix} \frac{1}{\sqrt{2}} & \frac{1}{\sqrt{2}} \\ \frac{1}{\sqrt{2}} & -\frac{1}{\sqrt{2}} \end{bmatrix} & \begin{bmatrix} \frac{1}{\sqrt{2}} & \frac{1}{\sqrt{2}} \\ \frac{1}{\sqrt{2}} & -\frac{1}{\sqrt{2}} \end{bmatrix} \\ \begin{bmatrix} \frac{1}{\sqrt{2}} & \frac{1}{\sqrt{2}} \\ \frac{1}{\sqrt{2}} & -\frac{1}{\sqrt{2}} \end{bmatrix} & -\begin{bmatrix} \frac{1}{\sqrt{2}} & \frac{1}{\sqrt{2}} \\ \frac{1}{\sqrt{2}} & -\frac{1}{\sqrt{2}} \end{bmatrix} \end{bmatrix} = \frac{1}{\sqrt{2}}\begin{bmatrix} H & H \\ H & -H \end{bmatrix}.$$

This pattern continues. The matrix that corresponds to inputting three qubits and sending all three through Hadamard gates can be written using $H^{\otimes 2}$.

$$H^{\otimes 3} = \frac{1}{\sqrt{2}}\begin{bmatrix} H^{\otimes 2} & H^{\otimes 2} \\ H^{\otimes 2} & -H^{\otimes 2} \end{bmatrix} = \frac{1}{2\sqrt{2}}\begin{bmatrix} \begin{bmatrix} 1 & 1 & 1 & 1 \\ 1 & -1 & 1 & -1 \\ 1 & 1 & -1 & -1 \\ 1 & -1 & -1 & 1 \end{bmatrix} & \begin{bmatrix} 1 & 1 & 1 & 1 \\ 1 & -1 & 1 & -1 \\ 1 & 1 & -1 & -1 \\ 1 & -1 & -1 & 1 \end{bmatrix} \\ \begin{bmatrix} 1 & 1 & 1 & 1 \\ 1 & -1 & 1 & -1 \\ 1 & 1 & -1 & -1 \\ 1 & -1 & -1 & 1 \end{bmatrix} & -\begin{bmatrix} 1 & 1 & 1 & 1 \\ 1 & -1 & 1 & -1 \\ 1 & 1 & -1 & -1 \\ 1 & -1 & -1 & 1 \end{bmatrix} \end{bmatrix}$$

$$= \frac{1}{2\sqrt{2}} \begin{bmatrix} 1 & 1 & 1 & 1 & 1 & 1 & 1 & 1 \\ 1 & -1 & 1 & -1 & 1 & -1 & 1 & -1 \\ 1 & 1 & -1 & -1 & 1 & 1 & -1 & -1 \\ 1 & -1 & -1 & 1 & 1 & -1 & -1 & 1 \\ 1 & 1 & 1 & 1 & -1 & -1 & -1 & -1 \\ 1 & -1 & 1 & -1 & -1 & 1 & -1 & 1 \\ 1 & 1 & -1 & -1 & -1 & -1 & 1 & 1 \\ 1 & -1 & -1 & 1 & -1 & 1 & 1 & -1 \end{bmatrix}$$

As n increases these matrices quickly get large, but it is always true that

$$H^{\otimes n} = \frac{1}{\sqrt{2}} \begin{bmatrix} H^{\otimes(n-1)} & H^{\otimes(n-1)} \\ H^{\otimes(n-1)} & -H^{\otimes(n-1)} \end{bmatrix},$$

and this gives us a recursive formula that lets us quickly calculate them. These products of matrices telling us how to act on tensor products are called *Kronecker products*.

For Simon's algorithm we are going to need to study these matrices in some detail, but for our next algorithm the key observation is that the entries in the top row of each these matrices are all equal to one another; for $H^{\otimes n}$ they all equal $\left(1/\sqrt{2}\right)^n$.

The Deutsch-Jozsa Algorithm

Deutsch's algorithm looked at functions of one variable. You were given one of these and had to determine whether it was a constant or balanced function. The Deutsch-Jozsa problem is a generalization of this.

We now have functions of n variables. The inputs for each of these variables, as before, can be either 0 or 1. The output is either 0 or 1. We are told that our function is either constant—all the inputs get sent to 0, or all the inputs get sent to 1 — or it is balanced—half the inputs get sent to 0 and the other half to 1. If we are given one of these functions at random, how many function evaluations do we need to determine whether the function belongs to the constant group or to the balanced group?

To illustrate, we consider the case when $n = 3$. Our function takes three inputs, each of which can take two values. This means that there are 2^3, or 8, possible inputs:

(0,0,0), (0,0,1), (0,1,0), (0,1,1), (1,0,0), (1,0,1), (1,1,0), (1,1,1).

Classically, suppose we evaluate $f(0,0,0)$ and get the answer that $f(0,0,0) = 1$. We cannot deduce anything from this piece of information alone, so we ask for another function evaluation, say $f(0,0,1)$. If we get $f(0,0,1) = 0$, then we are done. We know that the function cannot be constant, so it must be balanced. On the other hand, if we get $f(0,0,1) = 1$, we cannot deduce anything from the two pieces of information we have. In the worst possible scenario, we could get the same answer for the first four questions and still not be able to answer the question. For example, from the fact that $f(0,0,0) = 1$, $f(0,0,1) = 1$, $f(0,1,0) = 1$, $f(0,1,1) = 1$ we cannot determine whether or not the function is balanced. We need to ask one more question. If the answer to the next question is also 1, then we know the function is constant. If the answer is 0, then we know the function is balanced.

This analysis works in general. Given a function of n variables, there will be 2^n possible input strings. In the best-case scenario we can obtain the answer with just two questions to the oracle, but in the worst case it will take us $2^{n-1} + 1$ questions. Since the $n - 1$ appears as an exponent, the function is exponential. In the worst case it requires an exponential number of inquiries to the oracle. The Deutsch-Jozsa algorithm is a quantum algorithm that just requires one question to the oracle, so the speedup is substantial!

The first step, as in all of these questions, is to describe the oracle. For each of the functions we need to construct an orthogonal matrix that captures the essence of the function. We just generalize our previous construction.

Given any function $f(x_0, x_1, \ldots, x_{n-1})$ that has n boolean inputs and has just one boolean output, we construct the gate F given by the following circuit, where the slashes with n on the top lines indicate that we have n wires in parallel.

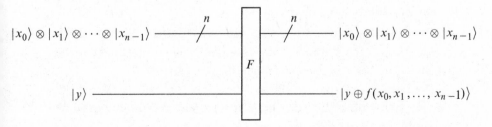

Remember that this circuit tells us what happens when each of the kets, $|x_i\rangle$, is either $|0\rangle$ or $|1\rangle$. The input consists of $n + 1$ kets, $|x_0\rangle \otimes |x_1\rangle \otimes \ldots \otimes |x_{n-1}\rangle$

and $|y\rangle$, where the first n correspond to the function variables. The output also consists of $n+1$ kets, the first n of which are exactly the same as the input kets. The last output is the ket $|f(x_0, x_1, \ldots, x_{n-1})\rangle$ if $y=0$ and the ket of the other boolean value when $y=1$.

The next step after describing how the black-box function works is to give the quantum circuit that incorporates this function. It is the natural generalization of the circuit used for Deutsch's algorithm: all the top qubits pass through Hadamard gates on either side of the black box.

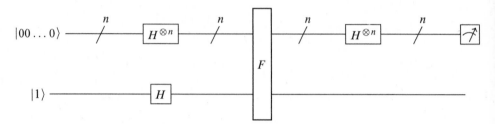

As before, we will analyze what this circuit does step by step. We show the case when $n=2$, just to make things look a little less messy on the page, but every step we do works in exactly the same way for every value of n.

Step 1. The Qubits Pass through the Hadamard Gates

The top n inputs are all $|0\rangle$. For $n=2$, this is $|00\rangle$. The following calculation shows what happens.

$$H^{\otimes 2}(|00\rangle) = \frac{1}{2}\begin{bmatrix} 1 & 1 & 1 & 1 \\ 1 & -1 & 1 & -1 \\ 1 & 1 & -1 & -1 \\ 1 & -1 & -1 & 1 \end{bmatrix}\begin{bmatrix} 1 \\ 0 \\ 0 \\ 0 \end{bmatrix} = \frac{1}{2}\begin{bmatrix} 1 \\ 1 \\ 1 \\ 1 \end{bmatrix} = \frac{1}{2}(|00\rangle + |01\rangle + |10\rangle + |11\rangle)$$

It gives a superposition of all possible states; each of the basis kets has the same probability amplitude (1/2 in this case).

(This calculation works for every value of n. After the n qubits have passed through $H^{\otimes n}$ they are in superposition of all possible states, each of which has the same probability amplitude: $\left(1/\sqrt{2}\right)^n$.)

The bottom entry is just $|1\rangle$. This becomes $\left(1/\sqrt{2}\right)|0\rangle - \left(1/\sqrt{2}\right)|1\rangle$ after passing through the Hadamard gate. At this stage, our three input qubits will be in the following state.

$\frac{1}{2}(|00\rangle + |01\rangle + |10\rangle + |11\rangle) \otimes \left(\frac{1}{\sqrt{2}}|0\rangle - \frac{1}{\sqrt{2}}|1\rangle\right)$. We will rewrite this as

$$\frac{1}{2\sqrt{2}}|00\rangle \otimes (|0\rangle - |1\rangle)$$

$$+ \frac{1}{2\sqrt{2}}|01\rangle \otimes (|0\rangle - |1\rangle)$$

$$+ \frac{1}{2\sqrt{2}}|10\rangle \otimes (|0\rangle - |1\rangle)$$

$$+ \frac{1}{2\sqrt{2}}|11\rangle \otimes (|0\rangle - |1\rangle)$$

Step 2. The Qubits Pass through the F Gate

After passing through the F gate the qubits will be in the following state.

$$\frac{1}{2\sqrt{2}}|00\rangle \otimes (|f(0,0)\rangle - |f(0,0) \oplus 1\rangle)$$

$$+ \frac{1}{2\sqrt{2}}|01\rangle \otimes (|f(0,1)\rangle - |f(0,1) \oplus 1\rangle)$$

$$+ \frac{1}{2\sqrt{2}}|10\rangle \otimes (|f(1,0)\rangle - |f(1,0) \oplus 1\rangle)$$

$$+ \frac{1}{2\sqrt{2}}|11\rangle \otimes (|f(1,1)\rangle - |f(1,1) \oplus 1\rangle)$$

We now use the fact that if a is either 0 or 1 we have the following

$$|a\rangle - |a \oplus 1\rangle = (-1)^a (|0\rangle - |1\rangle)$$

to rewrite the state as

$$(-1)^{f(0,0)} \frac{1}{2}|00\rangle \otimes \frac{1}{\sqrt{2}}(|0\rangle - |1\rangle)$$

$$+ (-1)^{f(0,1)} \frac{1}{2}|01\rangle \otimes \frac{1}{\sqrt{2}}(|0\rangle - |1\rangle)$$

$$+ (-1)^{f(1,0)} \frac{1}{2}|10\rangle \otimes \frac{1}{\sqrt{2}}(|0\rangle - |1\rangle)$$

$$+ (-1)^{f(1,1)} \frac{1}{2}|11\rangle \otimes \frac{1}{\sqrt{2}}(|0\rangle - |1\rangle)$$

As before, this shows that the bottom qubit is not entangled with the top qubits. We just look at the top two qubits. These top two are in state:

$$\frac{1}{2}\left((-1)^{f(0,0)}|00\rangle + (-1)^{f(0,1)}|01\rangle + (-1)^{f(1,0)}|10\rangle + (-1)^{f(1,1)}|11\rangle\right)$$

(The argument that we have used works for general n. At this stage you obtain a state that is a superposition of all basis kets. Each ket, $|x_0 x_1 \ldots x_{n-1}\rangle$ is multiplied by $\left(1/\sqrt{2}\right)^n (-1)^{f(x_0,x_1,\ldots,x_{n-1})}$.)

Step 3. The Top Qubits Pass through the Hadamard Gates

The standard method is to convert our state to a column vector and then multiply by the appropriate Kronecker product of the Hadamard matrix. This gives:

$$\frac{1}{4}\begin{bmatrix} 1 & 1 & 1 & 1 \\ 1 & -1 & 1 & -1 \\ 1 & 1 & -1 & -1 \\ 1 & -1 & -1 & 1 \end{bmatrix}\begin{bmatrix} (-1)^{f(0,0)} \\ (-1)^{f(0,1)} \\ (-1)^{f(1,0)} \\ (-1)^{f(1,1)} \end{bmatrix}.$$

However, we are not going to calculate all of the entries in the resulting column vector. We are just going to calculate the top entry. This entry comes from multiplying the bra corresponding to the top row of the matrix with the ket given by the column vector. We get

$$\frac{1}{4}\left((-1)^{f(0,0)} + (-1)^{f(0,1)} + (-1)^{f(1,0)} + (-1)^{f(1,1)}\right).$$

This is the probability amplitude of the ket $|00\rangle$. We calculate this amplitude for the possible functions.

If f is constant and sends everything to 0, the probability amplitude is 1.

If f is constant and sends everything to 1, the probability amplitude is −1.

For the balanced functions, the probability amplitude is 0.

Step 4. Measure the Top Qubits

When we measure the top qubits we will get one of 00, 01, 10, or 11. The question becomes "do we get 00?" If the function is constant, then we will with probability 1. If the function is balanced, we get it with probability 0. So, if the result of measuring gives 00, we know our function was constant. If the result is not 00, then the function is balanced.

The analysis works for general n. Just before we measure the qubits the probability amplitude for $|0\ldots0\rangle$ is

$$\frac{1}{2^n}\left((-1)^{f(0,0,\ldots,0)} + (-1)^{f(0,0,\ldots,1)} + \cdots + (-1)^{f(1,1,\ldots,1)}\right).$$

As with $n = 2$, this number will be ± 1 if f is constant and 0 if f is balanced. So, if every measurement gives 0, the function is constant. If at least one of the measurements is 1, the function is balanced.

Consequently, we can solve the Deutsch-Jozsa problem for any value n with just one use of the circuit. We only need to ask the oracle one question. Recall in the classical case, that in the worst case it required $2^{n-1} + 1$ questions, so the improvement is dramatic.

Simon's Algorithm

The two algorithms that we have seen so far have been unusual in that we get the final answer with certainty after just one query. Most quantum algorithms use a mixture of quantum algorithms and classical algorithms; they involve more than one use of a quantum circuit; and they involve probability. Simon's algorithm contains all of these components. However, before we describe the algorithm, we need to discuss the problem being tackled, and before we can do that, we need to introduce a new way of adding binary strings.

Bitwise Addition of Strings Modulo 2

We defined $\oplus$ to be the exclusive or *XOR*, or, equivalently, as addition modulo 2. Recall

$$0 \oplus 0 = 0 \quad 0 \oplus 1 = 1 \quad 1 \oplus 0 = 1 \quad 1 \oplus 1 = 0$$

We extend this definition to the addition of binary strings of the same length by the formula:

$$a_0 a_1 \cdots a_n \oplus b_0 b_1 \cdots b_n = c_0 c_1 \cdots c_n \text{ , where}$$

$$c_0 = a_0 \oplus b_0, c_1 = a_1 \oplus b_1, \ldots, c_n = a_n \oplus b_n .$$

This is like doing addition in binary, but ignoring any carries. Here's a concrete example of bitwise addition:

$$
\begin{array}{r}
1101 \\
\oplus \quad 0111 \\
\hline
1010
\end{array}
$$

The Statement of Simon's Problem

We have a function f that sends binary strings of length n to binary strings of length n. It has the property that there is some secret binary string s, such that $f(x) = f(y)$ if and only if $y = x$ or $y = x \oplus s$. We don't allow s to be the string consisting entirely of 0s; this forces pairs of distinct input strings to have the same output strings. The problem is to determine the secret string s. An example should make all of this clear.

We will take $n = 3$, so our function f will take binary strings of length 3 and give other binary strings of length 3. Suppose that the secret string is $s = 110$. Now

$$000 \oplus 110 = 110 \quad 001 \oplus 110 = 111 \quad 010 \oplus 110 = 100 \quad 011 \oplus 110 = 101$$
$$100 \oplus 110 = 010 \quad 101 \oplus 110 = 011 \quad 110 \oplus 110 = 000 \quad 111 \oplus 110 = 001\,.$$

Consequently, for this value of s, we get the following pairings:

$$f(000) = f(110) \quad f(001) = f(111) \quad f(010) = f(100) \quad f(011) = f(101).$$

A function with this property is

$$f(000) = f(110) = 101 \quad f(001) = f(111) = 010$$
$$f(010) = f(100) = 111 \quad f(011) = f(101) = 000\,.$$

Now, of course, we don't know the function f or the secret string s: We want to find s. The question is how many function evaluations need to be made to determine this string?

We keep evaluating the function f on strings. We stop as soon as we get a repeated answer. Once we have found two input strings that give the same output, we can immediately calculate s.

For example, if we find that $f(011) = f(101)$, then we know that

$$011 \oplus s_0 s_1 s_2 = 101.$$

Using the fact that

$$011 \oplus 011 = 000,$$

bitwise add 011 to the left of both sides of the equation to obtain

$$s_0 s_1 s_2 = 011 \oplus 101 = 110.$$

How many evaluations do we have to make using a classical algorithm? We have eight binary strings. It is possible to evaluate four of these and get four different answers, but with the fifth evaluation we are guaranteed to get a pair. In general, for strings of length n, there are 2^n binary strings and, in the worst case, we will need to $2^{n-1} + 1$ function evaluations to get a repeat. So, in the worst case, we will need to ask the oracle $2^{n-1} + 1$ questions.

Before we look at the quantum algorithm, we need to look at the Kronecker product of Hadamard matrices in a little more detail.

The Dot Product and the Hadamard Matrix

Given two binary strings, $a = a_0 a_1 \cdots a_{n-1}$ and $b = b_0 b_1 \cdots b_{n-1}$ of the same length, we define the *dot product* by

$a \cdot b = a_0 \times b_0 \oplus a_1 \times b_1 \oplus \cdots \oplus a_{n-1} \times b_{n-1}$, where $\times$ denotes our usual multiplication.

So, for example, if $a = 101$ and $b = 111$, then $a \cdot b = 1 \oplus 0 \oplus 1 = 0$. This operation can be thought of as multiplying corresponding terms of the sequences, then adding and finally determining whether the sum is odd or even.

In computer science, we often start counting at 0, so instead of counting from 1 to 4 we count from 0 to 3. Also, we often use binary. The numbers 0, 1, 2, and 3 are represented in binary by 00, 01, 10, 11. Given a 4×4 matrix, we will label both the rows and the columns with these numbers, as is shown here:

$$
\begin{array}{c}
 \begin{array}{cccc} 00 & 01 & 10 & 11 \end{array} \\
\begin{array}{c} 00 \\ 01 \\ 10 \\ 11 \end{array}
\left[
\begin{array}{cccc}
* & * & * & * \\
* & * & * & * \\
* & * & * & * \\
* & * & * & *
\end{array}
\right]
\end{array}
$$

The position of an entry in this matrix is given by listing both the row and the column in which it appears. If we make the entry in the ith row and jth column be $i \cdot j$, we get the following matrix.

$$
\begin{array}{c}
 \begin{array}{cccc} 00 & 01 & 10 & 11 \end{array} \\
\begin{array}{c} 00 \\ 01 \\ 10 \\ 11 \end{array}
\left[
\begin{array}{cccc}
0 & 0 & 0 & 0 \\
0 & 1 & 0 & 1 \\
0 & 0 & 1 & 1 \\
0 & 1 & 1 & 0
\end{array}
\right]
\end{array}
$$

Compare this matrix to $H^{\otimes 2}$. Notice that the entries that are 1 in our dot product matrix are in exactly the same positions as the negative entries in $H^{\otimes 2}$. Using the facts that $(-1)^0 = 1$ and $(-1)^1 = -1$, we can write

$$H^{\otimes 2} = \frac{1}{2}\begin{bmatrix} (-1)^{00\cdot 00} & (-1)^{00\cdot 01} & (-1)^{00\cdot 10} & (-1)^{00\cdot 11} \\ (-1)^{01\cdot 00} & (-1)^{01\cdot 01} & (-1)^{01\cdot 10} & (-1)^{00\cdot 11} \\ (-1)^{10\cdot 00} & (-1)^{10\cdot 01} & (-1)^{10\cdot 10} & (-1)^{10\cdot 11} \\ (-1)^{11\cdot 00} & (-1)^{11\cdot 01} & (-1)^{11\cdot 10} & (-1)^{11\cdot 11} \end{bmatrix}.$$

This method of finding where the positive and negative entries holds in general; for example, if we want the entry of $H^{\otimes 3}$ that is in the row with number 101 and column with number 111, we calculate the dot product and get 0. This tells us that the entry will be positive.

Hadamard Matrices and Simon's Problem

Now that we know how to find entries of Kronecker products of Hadamard matrices, we are going to use this knowledge to see what happens when we add two columns of one of these products. We are going to add two columns that are paired by the secret string s given in Simon's problem. If one column is labeled by the string b, the other will have label $b \oplus s$. We are going to add these two columns together.

To illustrate, we will work with strings of length 2, and suppose the secret string is 10. We will be adding column 00 to 10, or column 01 to 11.

Here is $H^{\otimes 2}$.

$$H^{\otimes 2} = \frac{1}{2}\begin{bmatrix} 1 & 1 & 1 & 1 \\ 1 & -1 & 1 & -1 \\ 1 & 1 & -1 & -1 \\ 1 & -1 & -1 & 1 \end{bmatrix}$$

Adding the 00 column to 10 gives:

$$\frac{1}{2}\begin{bmatrix} 1 \\ 1 \\ 1 \\ 1 \end{bmatrix} + \frac{1}{2}\begin{bmatrix} 1 \\ 1 \\ -1 \\ -1 \end{bmatrix} = \frac{1}{2}\begin{bmatrix} 2 \\ 2 \\ 0 \\ 0 \end{bmatrix}.$$

Adding the 01 column to 11 gives:

$$\frac{1}{2}\begin{bmatrix} 1 \\ -1 \\ 1 \\ -1 \end{bmatrix} + \frac{1}{2}\begin{bmatrix} 1 \\ -1 \\ -1 \\ 1 \end{bmatrix} = \frac{1}{2}\begin{bmatrix} 2 \\ -2 \\ 0 \\ 0 \end{bmatrix}.$$

Notice that some probability amplitudes are getting amplified and some are canceling. What exactly is going on here?

It is fairly easy to check that products and bitwise addition obey the usual law of exponents.

$$(-1)^{a\cdot(b\oplus s)} = (-1)^{a\cdot b}(-1)^{a\cdot s}.$$

This tells us that $(-1)^{a\cdot(b\oplus s)}$ and $(-1)^{a\cdot b}$ will be equal if $a\cdot s = 0$, and that $(-1)^{a\cdot(b\oplus s)}$ and $(-1)^{a\cdot b}$ will have opposite signs if $a\cdot s = 1$.

We can summarize this as:

$$(-1)^{a\cdot(b\oplus s)} + (-1)^{a\cdot b} = \pm 2 \quad \text{if} \quad a\cdot s = 0$$
$$(-1)^{a\cdot(b\oplus s)} + (-1)^{a\cdot b} = 0 \quad \text{if} \quad a\cdot s = 1$$

This tells us that when we add the two columns given by b and $b\oplus s$, the entry in row a will be 0 if $a\cdot s = 1$, and will be either 2 or -2 if $a\cdot s = 0$. In general, the entries cancel in the rows labeled with strings that have a dot product of 1 with the string s.

Looking back at our example, the reason that the bottom two entries are 0 is that these rows have labels 10 and 11 and these both have a dot product of 1 with our secret string s. The nonzero entries occur in the rows with labels 00 and 01 and these both have a dot product of 0 with s.

We now have the information needed to understand the quantum circuit for Simon's problem. It is going to give us a string whose dot product with the secret string s is 0. It is going to do that by adding two columns of the Hadamard matrix. Let's see how it works.

The Quantum Circuit for Simon's Problem

The first thing is to construct the black box—the gate that acts like f. The following circuit gives the construction.

We can think of this as inputting two strings consisting of $|0\rangle$s and $|1\rangle$s, both of which have the same length. The top string is unchanged. The bottom string is the function evaluated on the top string added bitwise to the bottom string.

The following circuit gives the circuit for the algorithm.

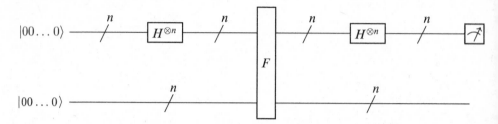

We will illustrate what happens in the case $n = 2$. Everything we do generalizes straightforwardly for general n.

The first step consists of the qubits in the top register passing through the Hadamard gates. This should now look familiar. The top two qubits are initially in state $|00\rangle$, while after passing through the Hadarmard gates they will be in state

$$\frac{1}{2}(|00\rangle + |01\rangle + |10\rangle + |11\rangle).$$

The bottom two qubits remain in state $|00\rangle$. So, at this stage the four qubits are in state:

$$\frac{1}{2}(|00\rangle \otimes |00\rangle + |01\rangle \otimes |00\rangle + |10\rangle \otimes |00\rangle + |11\rangle \otimes |00\rangle)$$

The next thing that happens is that the qubits pass through the F gate. This changes the state to:

$$\frac{1}{2}(|00\rangle \otimes |f(00)\rangle + |01\rangle \otimes |f(01)\rangle + |10\rangle \otimes |f(10)\rangle + |11\rangle \otimes |f(11)\rangle)$$

The top qubits now pass through the Hadamard gates, which changes the state to:

$$\frac{1}{4}(|00\rangle + |01\rangle + |10\rangle + |11\rangle) \otimes |f(00)\rangle$$

$$+ \frac{1}{4}(|00\rangle - |01\rangle + |10\rangle - |11\rangle) \otimes |f(01)\rangle$$

$$+ \frac{1}{4}(|00\rangle + |01\rangle - |10\rangle - |11\rangle) \otimes |f(10)\rangle$$

$$+ \frac{1}{4}(|00\rangle - |01\rangle - |10\rangle + |11\rangle) \otimes |f(11)\rangle$$

The pattern of + and − signs comes from the matrix for $H^{\otimes 2}$. We now rearrange the terms, solving for the first two qubits, which results in the following:

$$\frac{1}{4}|00\rangle \otimes (|f(00)\rangle + |f(01)\rangle + |f(10)\rangle + |f(11)\rangle)$$

$$+\frac{1}{4}|01\rangle \otimes (|f(00)\rangle - |f(01)\rangle + |f(10)\rangle - |f(11)\rangle)$$

$$+\frac{1}{4}|10\rangle \otimes (|f(00)\rangle + |f(01)\rangle - |f(10)\rangle - |f(11)\rangle)$$

$$+\frac{1}{4}|11\rangle \otimes (|f(00)\rangle - |f(01)\rangle - |f(10)\rangle + |f(11)\rangle)$$

This way of writing the state has a couple of nice features. The first is that here, also, the pattern of + and − signs comes from the matrix for $H^{\otimes 2}$. The second is that the pairs of qubits to the left of the tensor product correspond to the row numbers.

Now we use the fact that we know that $f(b) = f(b \oplus s)$, so $|f(b)\rangle = |f(b \oplus s)\rangle$. We can simplify things, combining these terms, by adding their probability amplitudes. This corresponds to the column addition we just looked at. To illustrate, suppose that $s = 10$, then $f(00) = f(10)$ and $f(01) = f(11)$. If we plug these values into the state, we obtain:

$$\frac{1}{4}|00\rangle \otimes (|f(00)\rangle + |f(01)\rangle + |f(00)\rangle + |f(01)\rangle)$$

$$+\frac{1}{4}|01\rangle \otimes (|f(00)\rangle - |f(01)\rangle + |f(00)\rangle - |f(01)\rangle)$$

$$+\frac{1}{4}|10\rangle \otimes (|f(00)\rangle + |f(01)\rangle - |f(00)\rangle - |f(01)\rangle)$$

$$+\frac{1}{4}|11\rangle \otimes (|f(00)\rangle - |f(01)\rangle - |f(00)\rangle + |f(01)\rangle)$$

which simplifies to

$$\frac{1}{4}|00\rangle \otimes (2|f(00)\rangle + 2|f(01)\rangle)$$

$$+\frac{1}{4}|01\rangle \otimes (2|f(00)\rangle - 2|f(01)\rangle)$$

$$+\frac{1}{4}|10\rangle \otimes (0)$$

$$+\frac{1}{4}|11\rangle \otimes (0)$$

The kets to the left of the tensor products are labeled with the row numbers of the matrix. The 0s on the right of the tensor products occur in the rows whose dot product with s is 1.

We can simplify the state to:

$$\frac{1}{\sqrt{2}}|00\rangle \otimes \frac{1}{\sqrt{2}}(|f(00)\rangle + |f(01)\rangle) + \frac{1}{\sqrt{2}}|01\rangle \otimes \frac{1}{\sqrt{2}}(|f(00)\rangle - |f(01)\rangle).$$

When we measure the top two qubits we will get either 00 or 01, each with probability 1/2.

Though we have only looked at the relatively simple case of $n = 2$, everything we have done holds for every value of n. At the end of the process we will end up with one of the strings whose dot product with the secret string is 0. Each of these strings is equally likely.

You might be concerned that after all this work we still don't know s. This is where the classical part of Simon's algorithm comes in.

The Classical Part of Simon's Algorithm

We start with an example with $n = 5$. We know that there is some secret number $s = s_0s_1s_2s_3s_4$. We are not allowing 00000, so there are $2 \wedge 5 - 1 = 31$ possible choices for s. We are going to try to find it using Simon's quantum circuit.

We run it and get 10100 as an answer. We know that the dot product of this with s gives 0. So,

$$1 \times s_0 \oplus 0 \times s_1 \oplus 1 \times s_2 \oplus 0 \times s_3 \oplus 0 \times s_4 = 0.$$

This tells us that $s_0 \oplus s_2 = 0$. Since these digits are either 0 or 1, we deduce that $s_0 = s_2$.

We run the circuit again hoping that we don't get 10100 again. (The probability of this happening is $1/16$, so we are fairly safe.) We also hope that we don't get 00000, which wouldn't give us any new information. Suppose we get 00100. Then we know that

$$0 \times s_0 \oplus 0 \times s_1 \oplus 1 \times s_2 \oplus 0 \times s_3 \oplus 0 \times s_4 = 0.$$

This shows that s_2 must be 0. From the first step, we can now deduce that s_0 must also be 0. We run the circuit again and get 11110. We know that

$$1 \times 0 \oplus 1 \times s_1 \oplus 1 \times 0 \oplus 1 \times s_3 \oplus 0 \times s_4 = 0,$$

which tells us that $s_1 = s_3$. Running the circuit again gives 00111, telling us that

$$0 \times 0 \oplus 0 \times s_1 \oplus 1 \times 0 \oplus 1 \times s_3 \oplus 1 \times s_4 = 0.$$

Consequently, we must have $s_3 = s_4$ and, since $s_1 = s_3$, we have $s_1 = s_3 = s_4$.

We know that not all of the digits are 0, so we must have $s_1 = s_3 = s_4 = 1$, and consequently s must be 01011. For this example, we made four calls to the oracle.

At this point, there are a couple of questions that you might be asking. The first concerns the algorithm for finding s using the outputs of the quantum circuit. We have seen what to do in a specific case, but is there an algorithm—a step-by-step procedure—that tells you what to do in every case? The second question concerns how we are measuring the number of questions we are asking the oracle. When we looked at the classical algorithm, we took the worst possible case and saw that after $2^{n-1} + 1$ questions we would definitely have our answer. But when we come to the quantum algorithm, the worst possible case is much worse! We are getting an answer at random. The answer does have a dot product of 0 with s, but we could get the same answer more than once. We could run our quantum circuit $2^{n-1} + 1$ times and get a string of 0s every single time. It is unlikely, but it is possible. A string of 0s gives us no information, so it is possible that after $2^{n-1} + 1$ questions to the oracle we haven't deduced anything at all about the secret number. We will address both of these concerns.

Each time we run the circuit, we get a number whose dot product with s is zero. This gives us a linear equation in the n unknowns. Running the circuit several times results in our obtaining several—a system of—equations. In the previous example, at each stage we got a new equation, but that new equation also gave us some new information. The technical term for this is that the equation is *linearly independent* of the previous equations. In order to calculate s we need a system of $n - 1$ linearly independent equations.***

Algorithms for solving systems of equations are extremely well known. They are studied in courses like Linear Algebra and Matrix Theory and have

---

*** You may have seen systems of linear equations before and remember that you need n equations to solve a system with n unknowns. This is true when the coefficients can be real numbers, but in our case the coefficients are either 0 or 1. This restriction and the fact that the string of all 0s is not allowed for s make it possible for us to reduce the number of equations by one.

numerous applications. They are so commonly needed that they are programmed into most scientific calculators. We won't discuss them here apart from mentioning that the number of steps required to solve a system of n equations can be bounded above by a quadratic expression involving n. We say the system can be solved in quadratic time.

The other question that we need to address is this: How many times do we need to run the quantum circuit? As we pointed out, in the worst-case scenario, we can keep running our qubits through the circuit and never get any useful information. However, this is extremely unlikely. We examine this idea in more detail in the next section.

Complexity Classes

In complexity theory, the main classification is between problems that take polynomial time to solve and those that need more than polynomial time. Polynomial time algorithms are regarded as being practical even for very large values of n, but non-polynomial time algorithms are regarded as being infeasible for large n.

Problems that classical algorithms can solve in polynomial time are denoted by P. Problems that quantum algorithms can solve in polynomial time are denoted by QP (sometimes it is denoted by EQP, for exact quantum polynomial time). Usually when we use these terms we are referring to the number of steps that an algorithm takes, but, remember, we defined a new way of measuring complexity—query complexity—that counts the number of questions we need to ask an oracle. We saw that the Deutsch-Jozsa problem was not in the class P, but belonged to QP for query complexity. (The constant function is a degree 0 polynomial.) This is sometimes described as saying that the Deutsch-Jozsa problem separates P and QP—it is a problem that belongs to QP but not to P for query complexity.

However, let's recall the worst-case scenario for the classical algorithm. To make things more concrete, we will take $n = 10$. We are given a function that takes 10 inputs and told that it is either balanced or constant. We have to keep evaluating our function on specific inputs until we can deduce the answer. There are $2^{10} = 1024$ possible inputs. The worst-case scenario is when the function is balanced, but we get the same answer for the first 512 evaluations, and then on the 513th evaluation we get the other value. But how likely is this to happen?

If the function is balanced, for each input value we are equally likely to get either a 0 or a 1. This can be compared to tossing a fair coin and obtaining a head or a tail. How likely is it to toss a fair coin 512 times and get heads every time? The answer is $(1/2)^{512}$, which is less than 1 divided by a googol, where a googol is 10^{100}. It is a minute number!

Suppose you were given a coin and asked whether it was fair or was double-headed. If you toss it once and get heads, you can't really answer the question. But if you toss it ten times and it comes up heads every time, then you can be fairly sure that it is double-headed. Of course, you could be wrong, but in practice we are willing to accept being wrong as long as the probability of this happening is very small.

This is what we do for the bounded-error complexity classes. We choose some bound on the probability of getting an error that we think is acceptable. Then we look at algorithms that can answer the question within our bound for error.

Returning to the Deutsch-Jozsa example, suppose that we want at least a 99.9 percent success rate, or equivalently an error rate of less than 0.1 percent. If a function is balanced the probability of evaluating the function 11 times and getting 0 every time is 0.00049 to five decimal places. Similarly, the probability of obtaining 1 every time is 0.00049. Consequently, the probability of obtaining the same answer 11 times in a row when the function is balanced is just less than 0.001. So if we are willing to accept a bound on the probability of error of 0.1 percent, we can choose to make at most 11 function evaluations. If during the process we get both a 0 and a 1, we can stop and know with certainty that our function is balanced. If all 11 evaluations are the same, we will say the function is constant. We could be wrong, but our error rate is less than our chosen bound. Notice that this argument works for any n. In every case, we need 11 function evaluations at most.

Problems that classical algorithms can solve in polynomial time with the probability of error within some bound are denoted *BPP* (for bounded-error probabilistic polynomial time). The Deutsch-Jozsa problem is in the class *BPP*.

One thing that you might be worried about is whether a problem could be in *BPP* for one bound on the probability of error, but not in the class *BPP* for a smaller bound. This doesn't happen. If the problem is in the class *BPP*, it will be there for every choice of the bound.

We now return to Simon's algorithm. We need to keep sending qubits through the circuit until we have $n-1$ linearly independent equations. As we know, in the worst case this process can go on forever, so Simon's algorithm is not in class QP. However, let's choose a bound that we are willing to accept on the probability of making an error. Then we can calculate N so that $(1/2)^N$ is less than our bound.

We won't prove this, but it can be shown that if we run the circuit $n+N$ times, the probability of the $n+N$ equations containing a system $n-1$ linearly independent equations is greater than $1-(1/2)^N$.

We can now state Simon's algorithm. First we decide on a bound on the probability of error and calculate the value N. Again, the number N does not depend on n. We can use the same value of N in each case. We run Simon's circuit $n+N$ times. The number of queries is $n+N$, which, since N is fixed, is a linear function of n. We make the assumption that our system of $n+N$ equations contains $n-1$ independent vectors. We could be wrong, but the probability of being wrong is less than the bound that we chose. Then we solve the system of $n+N$ equations using a classical algorithm. The time taken will be quadratic in $n+N$, but because N is a constant, this can be expressed as a quadratic in n.

The algorithm as a whole contains the quantum part that takes linear time added to the classical part that takes quadratic time, giving quadratic time overall. Problems that quantum algorithms can solve in polynomial time with the probability of error within some bound are denoted BQP (for bounded-error quantum polynomial time). Simon's algorithm shows the problem belongs to BQP for query complexity.

We showed that the classical algorithm, in the worst case, took $2^{n-1}+1$ function evaluations—this is exponential in n, not polynomial, so the problem definitely does not belong to P. It can also be shown that even if we allow a bound on the probability of error the algorithm is still exponential, so the problem does not belong to BPP. We say that Simon's problem separates BPP and BQP for query complexity.

Quantum Algorithms

We started this chapter by describing how in many popular descriptions the speedup provided by quantum algorithms is said to come solely from quantum parallelism—the fact that we can put the input into a superposition that involves all the basis states. However, we have looked at three algo-rithms and have seen that though we need to use quantum parallelism, we need to do much more. We will briefly look at what is needed and why it is hard!

The three algorithms we have studied are the most elementary and con-sidered standard, but as you have probably noticed they are by no means trivial. The dates when they were published tells an important story. David Deutsch published his algorithm in his landmark paper of 1985. This was the first quantum algorithm, and it showed that a quantum algorithm could be faster than a classical one. Deutsch and Jozsa published their gen-eralization of Deutsch's algorithm in 1992, seven years later. It might seem surprising that what seems to be a fairly straightforward generalization took so long to find, but it is important to realize that it is the modern notation and presentation that make the generalization seem to be the natural one. Deutsch's paper doesn't state the problem exactly the way it is stated here, and it doesn't use diagrams for quantum circuits that are now standard. That said, there was an incredibly productive period from 1993 to 1995 when many of the most important algorithms were discovered. Daniel Simon's algorithm was published in this window, as were the algorithms by Peter Shor and Lov Grover that we will look at in the next chapter.

Orthogonal matrices represent quantum gates. Quantum circuits con-sist of combinations of gates. These correspond to multiplying orthogonal matrices, and since the product of orthogonal matrices results in an orthog-onal matrix, any quantum circuit can be described by just one orthogonal matrix. As we have seen, an orthogonal matrix corresponds to a change of basis—a different way of viewing the problem. This is the key idea. Quan-tum computing gives us more ways of viewing a problem than classical computing does. But in order to be effective, there has to be a view that shows the correct answer separated from other possible incorrect answers. Problems that quantum computers can solve faster than classical comput-ers need to have a structure that becomes visible only when we transform it using an orthogonal matrix.

The problems that we have looked at are clearly reverse-engineered. They are not important problems that people have been considering for years and that we have only now discovered that if we view them from the right quantum computing perspective they become simpler to solve. Rather, they are problems that are specially created using the structure of

Kronecker products of Hadamard matrices. Of course, what we really want is not to reverse-engineer a problem, but to take a problem that is important and then construct a quantum algorithm that is faster than any known classical algorithm. This is what Peter Shor achieved in his 1994 landmark paper, in which he showed (among other things) how quantum computing could be used to break the codes that are currently used for Internet security. We will briefly discuss Shor's algorithm in the next chapter, where we look at the impact of quantum computing.

9 Impact of Quantum Computing

It is, of course, impossible to predict the long-term impact of quantum computing with any accuracy. If we look back at the birth of the modern computer in the 1950s, nobody could have predicted how much computers would change society and how dependent we would become on them. There are well-known quotes from computer pioneers proclaiming that the world would only need a handful of computers and that nobody would ever need a computer in their home. These quotes are out of context. The authors were generally referring to specific types of computers, but the impression they give, though exaggerated, is true. Initially computers were massive, had to be in air-conditioned rooms, and were not very reliable. Today, I have a laptop, a smartphone, and a tablet. All three are far more powerful than the first computers. I think that even visionaries like Alan Turing would be amazed at the extent to which computers have thoroughly permeated all levels of society. Turing did discuss chess playing and artificial intelligence, but nobody predicted that the rise of e-commerce and social media would come to dominate so much of our lives.

Quantum computing is now in its infancy, and the comparison to the first computers seems apt. The machines that have been constructed so far tend to be large and not very powerful, and they often involve superconductors that need to be cooled to extremely low temperatures. Already there are some people saying that there will be no need for many quantum computers to be built and that their impact on society will be minimal. But, in my opinion, these views are extremely shortsighted. Although it is impossible to predict what the world will be like in fifty years time, we can look at the dramatic changes in quantum computing over the last few years and see the direction in which it is heading. It might be some time before we get powerful universal quantum computers, but even before we do, quantum computing looks likely to make a substantial impact on our lives. In this chapter we will look at some ways that this could occur. In contrast to the previous chapter where we looked at three algorithms in considerable depth, we will look at a wide variety of topics at a less detailed level.

Shor's Algorithm and Cryptanalysis

The major result in quantum computing concerning cryptanalysis is Shor's algorithm. To fully understand this algorithm requires a substantial mathematics background. It uses Euler's theorem and continued fraction expansions from number theory. It also requires knowledge of complex analysis and the discrete Fourier transform. It marks the place where the theory of quantum computation changes from requiring just elementary mathematics to a more substantial background. Consequently, we won't be covering the algorithm in detail, but its importance means that we should at least look at it.

It is an algorithm, like Simon's algorithm, that has a quantum part and a classical part. The quantum part is similar to that of Simon's algorithm. Before we give a brief description, we will look at the problem that Shor wanted to tackle.

RSA Encryption

The RSA encryption method is named after its inventors, Ron Rivest, Adi Shamir, and Leonard Adleman. They published a paper on it and then patented it in 1978. Later it became known that Clifford Cocks, working for the Government Communications Headquarters (GCHQ), a British intelligence agency, had essentially invented the same algorithm in 1973. The British classified it, but they did pass it on to the Americans. It seems, however, that neither of the American or British intelligence agencies used it or realized how important it would become. Nowadays, it is used widely on the Internet for encrypting data sent from one computer to another. It is used for Internet banking and for electronic purchases using credit cards.

We will show how the encryption algorithm works with an example in which we want to share some confidential information with our bank and, at the same time, want to protect it from anyone who might be eavesdropping.

When you want to communicate with the bank, you want to encrypt your data so that if it is intercepted it cannot be read. The actual encryption of the data is going to be done using a key that both you and the bank share to both encrypt and decrypt—this is called a symmetric key—and must be kept secret by both parties. The key is generated on your computer and sent to the bank, but, of course, we can't just send the key without encrypting it. We need to encrypt the key that we are going to use to encrypt our communication with the bank. This is where RSA encryption enters the picture. It is a way of securely sending the key to the bank.

To start the communication with the bank, your computer generates the key that will be used later for encryption and decryption for both you and the bank. We will call the key K.

The bank's computer finds two large prime numbers that we will denote by p and q. The primes need to be roughly the same size and the product $N = pq$, called the modulus, should contain at least 300 digits using standard decimal numbers (1024 binary digits), which is currently considered large enough to ensure security. This is fairly straightforward. There are efficient ways of generating these primes and multiplying the two primes to get the modulus N is easy.

The second step is for the bank to find a relatively small number e that shares no common factors with either $p-1$ or $q-1$. This is also easy to compute. The bank keeps the primes p and q secret, but sends the numbers N and e.

Your computer takes the key K and raises it to the power e taking the remainder after dividing by N. Once more, this is easy to do. This is the number is called $K^e \bmod N$. This is then sent to the bank. The bank knows how to factor N into p and q, and this lets it quickly calculate K.

If someone is tapping into the communication, they will know N and e, both of which the bank sent, they will also know the number $K^e \bmod N$ that you sent. To calculate K, the eavesdropper needs to know the factors p and q of N, but these are being kept secret. The security of the system depends on the fact that the eavesdropper will not be able to factor the number N to obtain p and q.

The question is how hard is it to factor a number that is the product of two large primes? The answer is that it seems hard. All of the other steps involved in RSA encryption can be performed with classical algorithms that take polynomial time, but nobody has discovered a classical algorithm that can factor a product of two large primes in polynomial time. But, on the other hand, nobody has a proof that such an algorithm doesn't exist.

This is where Shor enters the picture. He constructed a quantum algorithm that does factor a product of large prime numbers. The algorithm belongs to class *BQP*, which means that it works with bounded error in polynomial time. One thing that needs to be emphasized is that we are no longer talking about query complexity. We are not assuming that we can ask questions of an oracle. We are counting the total number of steps or, equivalently, the time needed to get from the beginning to the end of the computation. Shor is giving a concrete algorithm for each step. The fact that the algorithm belongs to *BQP* means that if it is implemented it becomes feasible to factor large numbers, and, more important, it means that if the quantum circuit can be actually constructed, then RSA encryption is no longer secure.

Shor's Algorithm

Shor's algorithm involves a significant amount of mathematics. We will just give a short and somewhat vague description of the quantum part.

An important part of the algorithm is a gate that is called the *quantum Fourier transform gate*. This can be thought of as a generalization of the Hadamard gate. In fact, for one qubit the quantum Fourier transform gate is exactly H. Recall that we used a recursive formula that told us how to get from the matrix for $H^{\otimes n-1}$ to the matrix for $H^{\otimes n}$. Similarly, we can give a recursive formula for the quantum Fourier transform matrix. The major difference between $H^{\otimes n}$ and the quantum Fourier matrix is that the entries in the latter case are generally complex numbers—more specifically, they are complex roots of unity. Recall that the entries for $H^{\otimes n}$ are either 1 or -1. These are the two possible square roots of 1. When we look for fourth roots of 1 we again just get ± 1 if we are using real numbers, but we get two other roots if we use the complex numbers. In general, 1 has n complex nth roots. The quantum Fourier transform matrix on n qubits involves all the 2^nth complex roots of unity.

Simon's algorithm was based on the properties of $H^{\otimes n}$. It used interference, the amplitudes were either 1 or -1, which meant that when we added terms, the kets either canceled or reinforced one another. Shor realized that a similar idea applied to the quantum Fourier matrix, only now the amplitudes are given not just by 1 and -1, but also by all the 2^nth complex roots of unity, which means that we can detect more types of periods than just the ones that Simon's algorithm considers.

Recall that we know the number N and want to factor it into the product of the two primes p and q. The algorithm chooses a number a satisfying $1 < a < N$. It checks to see if a shares any factors with N, if it does we can deduce that a is a multiple of either p or q. From there it is easy to complete the factorization. If a does not share any factors with N, then we calculate $a \bmod(N)$, $a^2 \bmod(N)$, $a^3 \bmod(N)$, and so on, where $a^i \bmod(N)$ means calculate a^i and then take the remainder when divided by N. Since these numbers are remainders, they will all be less than N. Consequently, this sequence of numbers will eventually repeat. There will be some number r such that $a^r \bmod(N) = a \bmod(N)$. The number r can be thought of as the period, and it is this number that the quantum part of Shor's algorithm computes. Once r has been found, classical algorithms can use this fact to determine the factors of N.

Well, that description was rather sketchy, but it gives some idea of how the quantum part of Shor's algorithm works. The key part is that Simon's algorithm for finding the secret string s can be generalized to find the unknown period r.

The algorithm actually has been implemented, but just for small numbers. In 2001, it was used to factor 15 and in 2012 it factored 21. Clearly, it is nowhere near factoring 300 digit numbers at the moment. But how long will it take before a circuit can be built for numbers of this size? It seems to be only a matter of time until the RSA encryption scheme will no longer be secure.

Over the years other methods of encryption have been developed, but Shor's algorithm also works on many of these. It has become clear that we need to develop new cryptographic methods—and these new methods should be able to withstand not just classical attacks but also attacks by quantum computers.

Post-quantum cryptography is now an extremely active area, with new methods of encryption being developed. Of course, there is no reason why these have to use quantum computing. We just need the encrypted message to be able to withstand being broken by a quantum computer. But quantum ideas do give us ways of constructing secure codes.

We have seen two quantum key distribution (QKD) schemes that are secure: the BB84 and Ekert's protocols. Several labs have succeeded in getting QKD systems up and running. There are also a few companies that offer QKD systems for sale. One of the first times that QKD was used in a real-world setting was in 2007, when ID Quantique set up a system to secure the transmission of votes between a counting station and Geneva's main polling office during a Swiss parliamentary election.

Many countries are experimenting with small quantum networks using optic fiber. There is the potential of connecting these via satellite and being able to form a worldwide quantum network. This work is of great interest to financial institutions.

The most impressive results, so far, involve a Chinese satellite that is devoted to quantum experiments. It's named Micius after a Chinese philosopher who did work in optics. This is the satellite that was used for the quantum teleportation we mentioned in an earlier chapter. It has also been used for QKD. A team in China connected to a team in Austria—the first time that intercontinental QKD has been achieved. Once the connection was secured, the teams sent pictures to one another. The Chinese team sent the Austrians a picture of Micius, and the Austrians sent a picture of Schrödinger to the Chinese.

Grover's Algorithm and Searching Data

We are entering the era of big data. Searching through enormous data sets efficiently is now a high priority for many major companies. Grover's algorithm has the potential to speed up data searches.

Lov Grover invented the algorithm in 1996. Like Deutsch's and Simon's algorithms, its speedup over classical algorithms is given in terms of query complexity. Of course, to implement the algorithm for real-world data searches, we don't have oracles that can answer our questions. We have to construct an algorithm that does the work of the oracle. But before we begin to discuss how to implement Grover's algorithm, we will look at what it does and how it does it.

Grover's Algorithm

Imagine that you have four cards in front of you. They are all face down. You know that one of them is the ace of hearts, and this is the card you want to find. How many cards must you turn over until you know the location of the ace of hearts?

You might be lucky and turn it over on the first try, or you might be unlucky and turn over three cards, none of which is the ace. If you are unlucky and haven't turned it over after three tries, then you know that the last card must be the ace. So, we know where the ace is after turning over between one and three cards. On average we have to turn over 2.25 cards.

This problem is one that Grover's algorithm tackles. Before we begin describing the algorithm, we will reword the problem. We have four binary strings: 00, 01, 10, and 11. We have a function f that sends three of these strings to 0 and the other one to 1. We want to find the find the binary string that is sent to 1. For example, we might have $f(00) = 0$, $f(01) = 0$, $f(10) = 1$, and $f(11) = 0$. The problem now asks how many function evaluations do we need to make before we find that $f(10) = 1$. We are just restating the problem, wording it in terms of functions instead of cards, so we know the answer is the same as before: 2.25 on average.

As with all query complexity algorithms we construct an oracle—a gate that encapsulates the function. For our example, where we just have four binary strings, the oracle is given in figure 9.1.

The circuit for Grover's algorithm is given in figure 9.2.

The algorithm has two steps. The first is to flip the sign of the probability amplitude connected to the location we are trying to find. The second is to amplify this probability amplitude. We will show how the circuit does this.

After going through the Hadamard gates, the top two qubits will be in state

$$\frac{1}{2}(|00\rangle + |01\rangle + |10\rangle + |11\rangle)$$

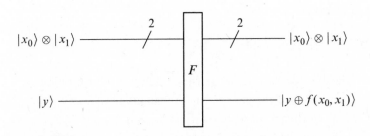

Figure 9.1
The oracle for f.

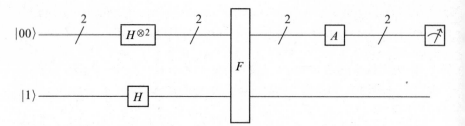

Figure 9.2
Grover algorithm circuit.

and the bottom qubit will be in state

$$\frac{1}{\sqrt{2}}|0\rangle - \frac{1}{\sqrt{2}}|1\rangle.$$

We can write the combined state as

$$\frac{1}{2}\left(|00\rangle \otimes \left(\frac{1}{\sqrt{2}}|0\rangle - \frac{1}{\sqrt{2}}|1\rangle\right) + |01\rangle \otimes \left(\frac{1}{\sqrt{2}}|0\rangle - \frac{1}{\sqrt{2}}|1\rangle\right)\right.$$

$$\left. + |10\rangle \otimes \left(\frac{1}{\sqrt{2}}|0\rangle - \frac{1}{\sqrt{2}}|1\rangle\right) + |11\rangle \otimes \left(\frac{1}{\sqrt{2}}|0\rangle - \frac{1}{\sqrt{2}}|1\rangle\right)\right).$$

The qubits then pass through the F gate. This flips $|0\rangle$ and $|1\rangle$ of the third qubit in the location we are trying to find. If we use our example where $f(10) = 1$, we obtain

$$\frac{1}{2}\left(|00\rangle \otimes \left(\frac{1}{\sqrt{2}}|0\rangle - \frac{1}{\sqrt{2}}|1\rangle\right) + |01\rangle \otimes \left(\frac{1}{\sqrt{2}}|0\rangle - \frac{1}{\sqrt{2}}|1\rangle\right)\right.$$

$$\left. + |10\rangle \otimes \left(\frac{1}{\sqrt{2}}|1\rangle - \frac{1}{\sqrt{2}}|0\rangle\right) + |11\rangle \otimes \left(\frac{1}{\sqrt{2}}|0\rangle - \frac{1}{\sqrt{2}}|1\rangle\right)\right).$$

This can be written as

$$\frac{1}{2}(|00\rangle + |01\rangle - |10\rangle + |11\rangle) \otimes \left(\frac{1}{\sqrt{2}}|0\rangle - \frac{1}{\sqrt{2}}|1\rangle\right).$$

The result is that the top two qubits are not entangled with the bottom qubit, but we have flipped the sign of the probability amplitude of $|10\rangle$, which corresponds to the location we are trying to find.

At this stage, if we were to measure the top two qubits we would get one of the four locations, with each of the four answers being equally likely. We need another trick and that is amplitude amplification. Amplitude amplification works by flipping a sequence of numbers about their mean. If a number is above the mean, it is flipped below the mean. If a number is below the mean, it is flipped above the mean. In each case the distance to the mean is preserved. To illustrate, we use the four numbers 1, 1, 1, and −1. Their sum is 2, and so their mean is 2/4, which is equal to 1/2. We then go through the sequence of numbers. The first is 1. This is 1/2 above the mean. When we flip about the mean it becomes 1/2 below the mean. In this case it becomes 0. The number −1 is 3/2 below the mean. When we flip about the mean it becomes 3/2 above the mean, which is 2.

Our top two qubits are currently in the state

$$\frac{1}{2}|00\rangle + \frac{1}{2}|01\rangle - \frac{1}{2}|10\rangle + \frac{1}{2}|11\rangle.$$

If we flip the probability amplitudes about the mean we get $0|00\rangle + 0|01\rangle + 1|10\rangle + 0|11\rangle = |10\rangle$. When we measure this we will get 10 with certainty, so flipping about the mean does exactly what we want. We just need to make sure that there is a gate or, equivalently, an orthogonal matrix that performs the flip about the mean. There is. It's

$$A = \frac{1}{2}\begin{bmatrix} -1 & 1 & 1 & 1 \\ 1 & -1 & 1 & 1 \\ 1 & 1 & -1 & 1 \\ 1 & 1 & 1 & -1 \end{bmatrix}.$$

When this gate acts on the top two qubits we get

$$A\left(\frac{1}{2}|00\rangle + \frac{1}{2}|01\rangle - \frac{1}{2}|10\rangle + \frac{1}{2}|11\rangle\right) = \frac{1}{4}\begin{bmatrix} -1 & 1 & 1 & 1 \\ 1 & -1 & 1 & 1 \\ 1 & 1 & -1 & 1 \\ 1 & 1 & 1 & -1 \end{bmatrix}\begin{bmatrix} 1 \\ 1 \\ -1 \\ 1 \end{bmatrix} = \begin{bmatrix} 0 \\ 0 \\ 1 \\ 0 \end{bmatrix} = |10\rangle.$$

For this example, where we just have two qubits, we only need to use the oracle once. We only need to ask one question. So, for $n = 2$, Grover's algorithm gives the answer with certainty with just one question, whereas the classical case takes 2.25 questions on average.

Exactly the same idea works for n qubits. We start by flipping the sign of the probability amplitude that corresponds to the location we are trying to find. Then we flip about the mean. However, the amplitude magnification is not as dramatic in general as in the case for just two qubits. For example, if we have eight numbers, seven of which are 1 and the other of which is −1. Their sum is 6, and so their mean is 6/8. When we flip about the mean, the 1s become 1/2s, and the −1 becomes 10/4. The consequence of this is that if we have three qubits, after performing the amplitude magnification, if we were to measure our qubits, we would get the location we are trying to find with higher probability than the other locations. The concern is that there is still some significant probability that we will get the wrong answer. We want a higher probability of getting the right answer—we want to magnify the amplitude even more before we measure. The solution is that we send everything back through the circuit. We flip the sign of the probability amplitude associated with the location we are trying to find again and then perform the flip about the mean again.

Let's look at the general case. We want to find something that could be in one of m possible locations. To find it classically we need to ask $m-1$ questions in the worst-case scenario. The number of questions grows at the same rate as the size of m. Grover calculated a formula for the number of times you should use his circuit to maximize the chance of getting the correct answer. The number given by this formula grows at the rate $\sqrt{m}$. This is a quadratic speedup.

Applications of Grover's Algorithm

There are a number of problems with implementing the algorithm. The first is that the quadratic speedup is for query complexity. If we need to use an oracle, then we need to actually construct it, and if we are not careful the number of steps involved with the computation of the oracle will outweigh the number of steps saved by the algorithm, resulting in the algorithm being slower rather than faster than the classical one. Another problem is that in calculating the speedup we are assuming that there is no underlying order to the data set. If there is structure, we can often find classical algorithms that exploit the structure and find the solution more quickly than randomly guessing. The last concern is about the speedup. Quadratic speedup is nothing like the exponential speedup we have seen with other algorithms. Can't we do better? Let's look at these concerns.

The concerns involving implementing the oracle and the structure of data sets are both valid and show that Grover's algorithm is not going to be practical for most database searches. But in certain cases the structure of the data can make it possible to construct an oracle that works efficiently. In these cases, the algorithm can give a speedup over classical algorithms. The question about whether we can do better than quadratic speedup has been answered. It has been proved that Grover's algorithm is optimal. There is no quantum algorithm that can solve the problem with more than just quadratic speedup. Quadratic speedup, though not as impressive as exponential speedup, is useful. With massive data sets, any speedup can be valuable.

The main applications for Grover's algorithm are probably not going to be for the algorithm as has been presented but for variations on it. In particular, the idea of amplitude amplification is a useful one.

We have only presented a few algorithms, but Shor's and Grover's are considered the most important. Many other algorithms have built upon the ideas in these two.* We now turn our attention from algorithms to other applications of quantum computing.

Chemistry and Simulation

In 1929, Paul Dirac wrote about quantum mechanics, saying, "The fundamental laws necessary for the mathematical treatment of a large part of physics and the whole of chemistry are thus completely known, and the difficulty lies only in the fact that the application of these laws leads to equations that are too complex to be solved."

In theory, all of chemistry involves interactions of atoms and configurations of electrons. We know the underlying mathematics—it's quantum mechanics, but although we can write down the equations we cannot solve them exactly. In practice, chemists use approximation techniques instead of trying to find exact solutions. These approximations ignore fine details. Computational chemistry has taken this approach and, in general, it has worked well. Classical computers can give us good answers in many cases, but there are areas where the current computational techniques don't work. The approximation is not good enough. You need the fine details.

* The online Quantum Algorithm Zoo, found at https://math.nist.gov/quantum/ zoo/, aims to provide a comprehensive catalog of all quantum algorithms.

Feynman thought that one of the main applications of quantum computers would be to simulate quantum systems. Using quantum computers to study chemistry that belongs to the quantum world is a natural idea that has great potential. There are a number of areas where it is hoped that quantum computing will make important contributions. One of these is to understand how an enzyme, nitrogenase, used to make fertilizers actually works. The current method of producing fertilizers releases a significant amount of greenhouse gases and consumes considerable energy. Quantum computers could play a major role in understanding this and other catalytic reactions.

There is a group at the University of Chicago that is looking into photosynthesis. The transfer of sunlight to chemical energy is a process that happens quickly and very efficiently. It is a quantum mechanical process. The long-term goal is to understand this process and then use it in photovoltaic cells.

Superconductivity and magnetism are quantum mechanical phenomena. Quantum computers may help us understand them better. One goal is to develop superconductors that don't need to be cooled to near absolute zero.

The actual construction of quantum computers is in its infancy, but even with a few qubits it is possible to begin studying chemistry. IBM recently simulated the molecule beryllium hydride (BeH_2) on a seven-qubit quantum processor. This is a relatively small molecule with just three atoms. The simulation does not use the approximations that are used in the classical computational approach. However, since IBM's processor uses just a few qubits, it is possible to simulate the quantum processor using a classical computer. Consequently, everything that can be done on this quantum processor can be done classically. However, as processors incorporate more qubits we get to the point where it is no longer possible to simulate them classically. We will soon be entering a new era when quantum simulations are beyond the power of any classical computer.

Now that we have seen some of the possible applications, we will briefly survey some of the ways that are being used to build quantum computers.

Hardware

To actually make practical quantum computers you need to solve a number of problems, the most serious being decoherence—the problem of your qubit interacting with something from the environment that is not part of the computation. You need to set a qubit to an initial state and keep it in that state until you need to use it. You also need to be able to construct gates and circuits. What makes a good qubit?

Photons have the useful properties of being easy to initialize and easy to entangle, and they don't interact very much with the environment, so they stay coherent for long times. On the other hand, it is difficult to store photons and have them ready when you need them. The properties of photons make them ideal for communication, but they are more problematic for building quantum circuits.

We have often used electron spin as an example. Can this be used? Earlier we mentioned the apparatus used in the loophole-free Bell test. It used electrons trapped in synthetic diamonds. These are manipulated by shining lasers on them. The problem has been scaling. You can construct one or two qubits but, at the moment, it is not possible to generate large numbers. Instead of using electrons, spins of the nucleus have also been tried, but scalability is again the problem.

Another method uses the energy levels of ions. Ion-trap computing uses ions that are held in position by electromagnetic fields. To keep the ions trapped vibrations must be minimized; cooling everything to near absolute zero does this. The ions' energy levels encode the qubits and lasers can manipulate these. David Wineland used ion traps to construct the first *CNOT* gate in 1995, for which he received a Nobel Prize, and in 2016 researchers at NIST entangled more than 200 beryllium ions. Ion-traps do have potential to be used in future quantum computers, but a number of computers are being constructed using a different approach.

To minimize the interaction of quantum computers with the environment, they are always protected from light and heat. They are shielded against electromagnetic radiation, and they are cooled. One thing that can happen in cold places is that certain materials become superconductors—they lose all electrical resistance—and superconductors have quantum properties that can be exploited. These involve things called Cooper pairs and Josephson junctions.

The electrons in a superconductor pair up, forming what are called *Cooper pairs*. These pairs of electrons act like individual particles. If you sandwich thin layers of a superconductor between thin layers of an insulator, you obtain a *Josephson junction*.** These junctions are now used in physics and engineering to create sensitive instruments for measuring magnetic fields. For our purposes, the important fact is that the energy levels of the Cooper pairs in a superconducting loop that contains a Josephson junction are discrete and can be used to encode qubits.

IBM uses superconducting qubits in its quantum computers. In 2016, IBM introduced a five-qubit processor that they have made available to everyone for free on the cloud. Anyone can design their own quantum circuit, as long as it uses five or fewer qubits, and run it on this computer. IBM's aim is to introduce quantum computing to a wide audience—circuits for superdense coding, for Bell's inequality, and a model of the hydrogen atom have all been run on this machine. A primitive version of Battleships has also been run, giving the coder the claim of constructing the first quantum computer multiplayer game. At the end of 2017, IBM connected a twenty-qubit computer to the cloud. This time it is not for education, but it is a commercial venture where companies can buy access.

Google is working on its quantum computer. It also uses superconducting qubits. Google is expected to announce in the near future that it has a computer that uses 72 qubits. What is special about this number?

Classical computers can simulate quantum computers if the quantum computer doesn't have too many qubits, but as the number of qubits increases we reach the point where that is no longer possible. Google is expected to announce that it has reached or exceeded this number, giving them the right to claim quantum supremacy—the first time an algorithm has been run on a quantum computer that is impossible to run, or simulate, on a classical computer. IBM, however, is not giving up without a fight. Its team, using some innovative ideas, has recently found a way to simulate a 56-qubit system classically, increasing the lower bound on the number of qubits needed for quantum supremacy.

As work continues on building quantum computers, we are likely to see spinoffs into other areas. Qubits, however we encode them, are sensitive to interactions with their surroundings. As we understand these interactions better we will be able to build better shields to protect our qubits, but we will also be able to design ways our qubits can measure their surroundings.

** Brian David Josephson received the Nobel Prize in physics for his work on how Cooper pairs can flow through a Josephson junction by quantum tunneling.

An example involves electrons trapped in synthetic diamonds. These are very sensitive to magnetic fields. NVision Imaging Technologies is a startup that is using this idea to build NMR machines that they hope will be better, faster, and cheaper than current ones.

Quantum Annealing

D-Wave has computers for sale. Their latest, the D-Wave 2000Q has, as you might guess from its name, 2,000 qubits. However, their computers are not general purpose, they are designed for solving certain optimization problems using quantum annealing. We will give a brief description of this.

Blacksmiths often need to hammer metal and bend metal. In the process, it can become hardened—various stresses and deformities occur in the crystal structure—making it hard to work. Traditional annealing is a method of restoring the uniform crystal structure, making the metal malleable once more. It's done by heating the piece of metal to a high temperature and then letting it slowly cool.

Simulated annealing is a standard technique, based on annealing, that can be used for solving certain optimization problems. For example, suppose we have the graph given in figure 9.3 and want to find the lowest point—the absolute minimum. Think of the graph as being the bottom of a two-dimensional bucket. We drop a ball bearing into the bucket. It will settle at the bottom of one of the valleys. These are labeled A, B, and C in the figure. We want to find C. The ball bearing may not land at the bottom

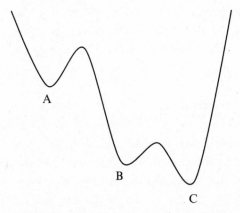

Figure 9.3
Graph of function—bottom of bucket.

of C, but instead it might end up at the bottom of valley at A. The important observation in annealing is that the energy required to push the ball bearing up the hill and let it drop into valley B is much less than the energy needed to push the ball bearing up from B and let it drop into A. So, we shake the bucket with an energy level between these two values. The ball can move from A to B, but it cannot move back. After a while of shaking at this level, it will end up either at the bottom of A or B. But shaking at this level can send the ball from C to B. The next step is to shake it again, but less energetically, with enough energy to get it up the peak from B to C, but not enough to let it get back from C to B.

In practice, you start shaking and gradually reduce the energy. This corresponds to gradually cooling your piece of metal in traditional annealing. The result is that the ball bearing ends up at the lowest point. You have found the absolute minimum of the function.

Quantum annealing adds quantum tunneling. This is a quantum effect where the ball bearing can just appear on the other side of a hill. Instead of going over, it can go through. Instead of reducing the heights of hills the ball can climb, you reduce the length of the tunnels it can tunnel through.

D-Wave has produced a number of commercially available computers that use quantum annealing for optimization problems. Initially, they were met with some skepticism about whether the computers actually used quantum tunneling, but now it is generally agreed that they do. There is still some question of whether the computers are faster than classical ones, but people are buying. Volkswagen, Google, and Lockheed Martin, among others, have all bought D-Wave machines.

After this brief look at hardware, we turn to deeper questions. What does quantum computation tell us about us, the universe, and what computation is at its most fundamental level?

Quantum Supremacy and Parallel Universes

There are 8 possible three-bit combinations: 000, 001, 010, 011, 100, 101, 110, 111. The number 8 comes from 2^3. There are two choices for the first bit, two for the second and two for the third, and we multiply these three 2s together. If instead of bits we switch to qubits, each of these 8 three-bit strings is associated with a basis vector, so the vector space is 8-dimensional. Exactly the same analysis tells us that if we have n qubits, then we will have

2^n basis vectors, and the space will be 2^n-dimensional. As the number of qubits grows, the number of basis vectors grows exponentially, and things quickly get big.

If we have 72 qubits, the number of basis elements is 2^{72}. This is about 4,000,000,000,000,000,000,000. It is a large number and is considered to be around the point at which classical computers cannot simulate quantum computers. Once quantum computers have more than 72 or so qubits we will enter the age of quantum supremacy—when quantum computers can do computations that are beyond the ability of any classical computer. As we mentioned earlier, it is expected that Google is about to announce that this age has been reached. (D-Wave has 2,000 qubits in its latest computer. However, this specialized machine has not been able to do anything that cannot be done by a conventional computer, so it hasn't broken the quantum supremacy barrier.)

Let's consider a machine with 300 qubits. This doesn't seem an unreasonable number for the not too distant future. But 2^{300} is an enormous number. It's more than the number of elementary particles in the known universe! A computation using 300 qubits would be working with 2^{300} basis elements. David Deutsch asks where computations like this, which involve more basis elements than there are particles in the universe, are done. He believes that we need to introduce parallel universes, each collaborating with one another.

This view of quantum mechanics and parallel universes goes back to Hugh Everett. Everett's idea is that, whenever we make a measurement, the universe splits into several copies, each containing a different outcome. Though this is distinctly a minority view, Deutsch is a firm believer. His paper in 1985 is one of the foundational papers in quantum computing, and one of Deutsch's goals with this work was to make a case for parallel universes. He hopes that one day that there will be a test, analogous to Bell's test, that will confirm this interpretation.

Computation

Alan Turing is one of the fathers of the theory of computation. In his landmark paper of 1936 he carefully thought about computation. He considered what humans did as they performed computations and broke it down to its most elemental level. He showed that a simple theoretical machine, which we now call a Turing machine, could carry out any algorithm. Turing's theoretical machines evolved into our modern day computers. They are universal computers. Turing's analysis showed us the most elemental operations. These involve the manipulation of bits. But remember, Turing was analyzing computation based on what humans do.

Fredkin, Feynman, and Deutsch argue that the universe does computations—that computations are part of physics. With quantum computation the focus changes from how humans compute to how the universe computes. Deutsch's 1985 paper should also be seen as a landmark paper in the theory of computation. In it, he showed that the fundamental object is not the bit, but the qubit.

We have seen that we will soon reach the point of quantum supremacy; that we will have quantum computers that no classical computers will be able to simulate, but what about the converse? Can quantum computers simulate classical computers? The answer is that they can. Any classical computation can be done on a quantum computer. Consequently, quantum computation is more general than classical computation. Quantum computations are not a strange way of doing a few special calculations; rather, they are a new way of thinking about computation as a concept. We shouldn't think of quantum and classical computation as two distinct subjects. Computation is really quantum computation. Classical computations are just special cases of quantum ones.

In this light, classical computation seems an anthropocentric version of what computation really is. Just as Copernicus showed that the Earth wasn't the center of the universe and Darwin showed that humans evolved from other animals, we are now beginning to see that computations are not centered on humans. Quantum computing represents a true paradigm shift.

I am not suggesting that classical computing is going to become obsolete, but it will become accepted that there is a more fundamental level of computing, and the most elemental level of computing involves qubits, entanglement, and superpositions. At the moment, the focus is on showing that certain quantum algorithms are faster than classical ones, but this will change. Quantum physics has been around longer than quantum computation. It's now accepted as its own subject. Physicists don't try to compare quantum physics with classical physics and hope to show that it is in some way better. They study quantum physics in its own right. The same shift will happen with quantum computation. We have been given new tools that change the way we study computation. We will use them to experiment and see what new things we can construct. This has started with teleportation and superdense coding, and it will continue.

We are entering a new era, with a new way of thinking about what computation really is. What we are going to discover is impossible to say, but now is the time for exploration and innovation. The greatest years for quantum computation are ahead of us.

Index